Washington Matthews, Jacob Lawson Wortman

The Human Bones of the Hemenway Collection

in the United States Army Medical Museum at Washington

Washington Matthews, Jacob Lawson Wortman

The Human Bones of the Hemenway Collection
in the United States Army Medical Museum at Washington

ISBN/EAN: 9783337218492

Printed in Europe, USA, Canada, Australia, Japan

Cover: Foto ©berggeist007 / pixelio.de

More available books at **www.hansebooks.com**

THE HUMAN BONES OF THE HEMENWAY COLLECTION IN THE UNITED STATES ARMY MEDICAL MUSEUM AT WASHINGTON,

BY

DR. WASHINGTON MATTHEWS,
SURGEON, U. S. ARMY;

WITH OBSERVATIONS ON THE HYOID BONES OF THIS COLLECTION

BY

DR. J. L. WORTMAN.

REPORTS PRESENTED TO THE NATIONAL ACADEMY OF SCIENCES, WITH THE APPROVAL OF THE SURGEON-GENERAL OF THE UNITED STATES ARMY,

BY

DR. JOHN S. BILLINGS,
SURGEON, U. S. ARMY.

In 1887 an expedition was fitted out under the direction of Mr. Frank Cushing, with funds supplied by the liberality of Mrs. Mary Hemenway, of Boston, for exploring certain ruins in the valley of the Gila River, in the Territory of Arizona.

The work of exploration was commenced with a mound of large size, apparently little more than a rude pile of earth, in the valley of the Salado, or Salt River, a tributary of the Gila. This proved to be the ruins of a large earthen house, apparently analogous in structure to the still standing Casa Grande, which lies about 35 miles to the southeast, and these ruins were found to be a part of a congregation of houses or a city, extending about 6 miles in length, and from half a mile to a mile in width along the valley. A large number of human bones were found under the floors of the houses, so large a number, in fact, that Mr. Cushing gave the place the name of Los Muertos, or the town of the dead.

When the work was fairly under way Mr. Cushing was taken sick, and application was made by the Hemenway Exploring Expedition to the Surgeon-General to allow Dr. Washington Matthews, of the Army, to go out and take Mr. Cushing's place during his illness, to supervise the explorations. Dr. Matthews went to Los Muertos in the month of August, 1887. He found that no attention had been paid to the collection or preservation of human bones, which were extremely fragile, crumbling to dust upon a touch, and which had been thrown about and trampled under foot by curious visitors, so that but little remained of value from the work which had been previously done. Recognizing the importance and interest of these remains, he set to work to preserve the bones excavated after his arrival as far as possible, and reported the facts to me, suggesting that, if possible, the anatomist of the Army Medical Museum, Dr. J. L. Wortman, should be sent out furnished with means for preserving these bones as fast as they were excavated, and carefully collecting and forwarding them to the Army Medical Museum for study.

In accordance with these suggestions Dr. Wortman went out in November, 1887, taking with him a supply of silicate of soda, glue, paraffin, and other materials for saturating and preserving the bones which should be discovered, and remained with the expedition, visiting several other localities, until June, 1888, when he returned to Washington.

141

The specimens of human bones thus obtained were carefully packed and forwarded to the Army Medical Museum, and after having been repaired and put in the best possible form, were examined and measured by Dr. Matthews, and his report of the results is herewith presented by authority of the Surgeon-General.

JOHN S. BILLINGS,
Surgeon, U. S. Army, Curator Army Medical Museum.

INTRODUCTION.

When we began the study of the bones described in this work we had reason to hope that a full general account of the expedition on which they were discovered, with its archæological labors and achievements, would be published simultaneously with, or in advance of, this report; in which case we should have embodied in this essay the results of our anthropometric studies only. But the continued illness of the director of the expedition, Mr. Frank Hamilton Cushing, has caused the indefinite postponement of the preparation of a general report, and we consequently have considered it advisable to present here a short introduction, setting forth the inception, objects, and results of that scholarly enterprise, short-lived, but fruitful in its results, which was known as the Hemenway Southwestern Archæological Expedition.

Along the great cordillera of the American Continent on both sides of the equator, through 75° of latitude, from Wyoming to Chile, extends a land abounding in ancient ruins.

A large part of this land of ruins lies within the boundaries of the United States. It contains the Territory of Arizona, most of Utah, more than half of New Mexico, extensive parts of the States of Colorado and Nevada, with small portions of Texas, and, perhaps, of California. Its precise boundaries are not known, for on its outskirts there is much wild and imperfectly explored country where the existence of ruins can neither be affirmed nor denied. Its approximate boundaries are: On the east, longitude 26° west (from Washington); on the west, longitude 38° west; on the north, latitude 41° north, and on the south the northern boundary of the republic of Mexico, 31.20° to 32° N. L. It covers about 400,000 square miles.

The great rivers which drain it into the ocean are the Colorado on the west and the Rio Grande on the east; the former flowing toward the Pacific, the latter toward the Atlantic. But much of the rain which falls on its surface does not reach the ocean; some is received in salty lakes which have no outlets; some goes to form streams which reach the great rivers only in seasons of abundant rain, but which at other times are absorbed by desert sands. It is an arid region, but not an absolute desert such as Gobi and Sahara. There is no part of it where rain does not fall some time during every year; but it is on the high mountains only that it descends abundantly; on the lower levels the precipitation of moisture is scanty, the dry seasons are long, and irrigation is essential to success in agriculture.

It has long been known that there were ruins in this arid region of the southwest. The earliest travelers, beginning with the Spanish conquerors of A. D. 1540, make mention of them, and their existence is noted in the reports of various military expeditions and public surveys which have entered this region since it was acquired by the United States from Mexico in 1848. The ruins have been known to the world for three centuries and a half; they have been in the possession of the United States for over forty years, yet it is only within the past four years (since April, 1887) that any attempt at systematic excavation has been made among them. In many of the better preserved ruins those portions which remained above the ground had been sketched, lithographed, photographed, engraved, surveyed, measured, modeled, and described, but the surface of the ground around and within them had not been broken. This method of examining them remained for the Hemenway Expedition to initiate.

The reasons for this tardiness on the part of our archæologists are numerous. This land of ruins was until recently wild, barren, and difficult of access; it was held largely by tribes of hostile Indians who to this day are not perfectly subdued. It is only within the last decade that

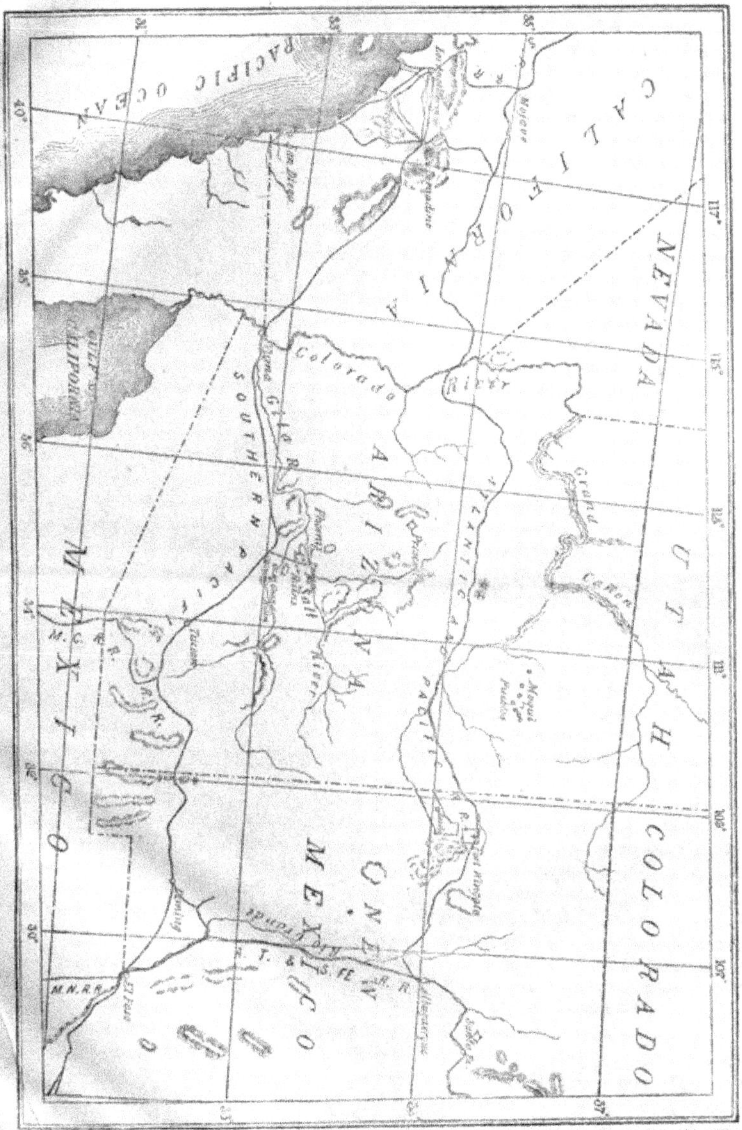

Fig. 1.—Map of southwestern portion of United States showing field of operations of the Hemenway Southwestern Archæological Expedition.

it has been crossed by railroads. Explorations within its borders were attended with many physical difficulties. The parties of topographical surveyors who entered the country had very short seasons in which to work, and they had neither the time nor means, had they had the inclination, to make the needed excavations. But besides physical hindrances there were others equally potent. The importance of excavation to the proper understanding of the archæology of this region was not appreciated; surface finds were numerous and interesting, and it was thought that excavation could yield nothing further. The majority of antiquarians in America were more deeply interested, as they still are, in the exploration of the old world than in that of the new. Money which was readily forthcoming for the one was withheld from the other by patrons of science in America.

The few explorers who were interested in work within our own borders found sufficient field for their labors and speculations in the mounds and kitchen-middens of the Eastern States. It was at length, through the unsatisfied curiosity of the ethnographer, not through the zeal of the archæologist, that the systematic exploration of the Western ruins was begun.

The region in question abounds in finely stratified sandstone, which with little labor may be prepared for building, and most of the ruins so far discovered are the remains of houses built of such stones. These may be found in all stages of decay—in some cases the walls are still standing many stories high, as in the valley of the Chaco; in other cases the sites are marked only by low heaps of lichen-covered stones, indistinguishable, save to the trained scientific eye, from natural accumulations of rocky *débris* with which the country abounds. Some of these ruins were

FIG. 2. The Casa Grande of the Gila.

inhabited by Indians within the brief historic period of New Mexico and Arizona, which extends over less than four centuries, but the vast majority are prehistoric. A number of the ruins are those of houses whose walls were of clay (*adobe* and a variety of *pisé*). Some of these in the valley of the Rio Grande were built since the Spanish occupation of the country and many have been erected under civilized guidance, but others, particularly those in the valley of the Colorado, are undoubtedly of prehistoric and aboriginal origin. As might be expected the earthen walls are in many cases reduced to the common level of the ground and are to be traced only, as in the ruined cities of the Salado, by digging beneath the surface of the earth; yet one of the best preserved and most imposing of the prehistoric ruins within our borders, the Casa Grande of the Gila (Fig. 2), is built of clay. This ruin was long supposed to be the remains of a structure without counterpart within the boundaries of the United States; but, as will hereafter be shown, it is now known to be but one of many such buildings which once towered over the wide flood-plains of the Gila and its tributaries.

In studying the folklore and religious practices of the people of Zuñi during his residence of about five years in their pueblo, Mr. Cushing found himself confronted by many perplexing questions for which no satisfactory explanation could then be found; but he was led to believe from the traditions of this people that some key to the problems might be discovered by exploring ruins far to the southwest of the Zuñi villages, where the people of Zuñi averred their ancestors once dwelled. We can not enter into a detailed account of these perplexing questions, nor can we relate how or why the explorer considers that he has solved them. It must be left for him to explain these matters fully at some future time.

EXPLORATIONS IN THE SALADO VALLEY.

It was not until the year 1886 that he found the pecuniary means to conduct the desired explorations, these being amply supplied by Mrs. Mary Hemenway, of Boston. Mr. Cushing set out with a party of assistants, to which others were afterwards added, and, in February, 1887, arrived in the neighborhood of the town of Tempe, in the valley of the Salado or Salt River, a tributary of the Gila, in the Territory of Arizona. Here he began by excavating some stone ruins on the rocky uplands, without any extraordinary results. While thus engaged his attention was attracted to certain earthen mounds situated on the level flood-plain of the Salado, and in particular to one of large size about 8 or 9 miles by road from Tempe. He proceeded to examine this mound and its vicinity.

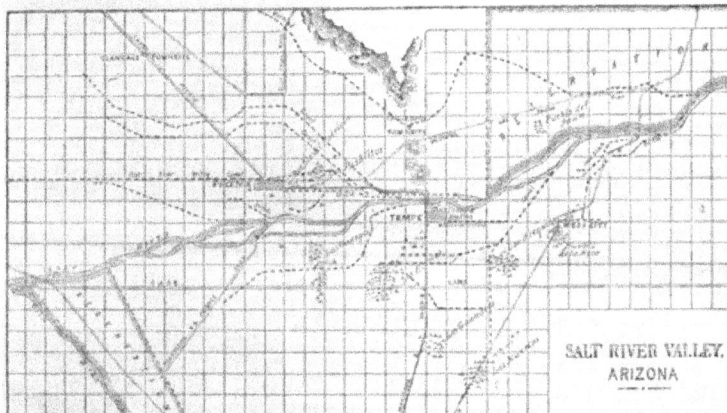

FIG. 5.—Map showing a part of the Salt River Valley, Maricopa County, Arizona, with modern towns, canals, and locations of ancient cities.

This mound seemed at first to be little more than a rude pile of earth. It had an irregular rectangular form, and had some appearance of being terraced. The surrounding level plain, covered with an abundant growth of that leguminous shrub or small tree, the mesquite (*Prosopis juliflora* D. C.), which is so common in the arid lands along our southwestern borders, presented to the untrained eye no remains of human habitation; but from fragments of pottery and other objects strewn over the ground, the explorer was led to believe that something of importance was hidden under the surface. He caused a trench to be dug and soon brought to light the foundations of earthen walls. Without delay he established his camp at this place and pursued his excavations with energy. The result was the discovery of an extensive collection of habitations—a city it might be called—some 6 miles in length and from half a mile to a mile in width. The mound proved to be the *débris* of a great earthen house of many stories and many chambers and analogous in structure to the still standing Casa Grande before referred to, which is distant from the mound to the southeast less than 35 miles in a direct line. In the course of excavation at this place so many skeletons were found under the floors of the houses that Mr. Cushing devised for it the Spanish name of Pueblo de los Muertos, or, briefly, Los Muertos, the town of the dead; and this name was retained for it, although he subsequently found other ruined cities in the vicinity where skeletons were as common as here.

S. Mis. 169——10

Work was continued in the valley of the Salado or Salt River until June 1888, a period of about sixteen months. During this time, besides isolated ruins and small groups of ruins, the party discovered the remains of six other large cities within a distance of about 10 miles from that first discovered. Of these, three were named: First, Las Acequias from the number, size, and distinct appearance in its vicinity of the old acequias or irrigating ditches through which the deported inhabitants conducted water to their fields; second, Los Hornos or The Ovens, from the number of earthen ovens found there, and third, Los Guanacos, because in it were found small terra-cotta images of animals thought to resemble the guanaco of South America. In these ruined cities the remains of other buildings like the Casa Grande were found.

HOUSES.

The houses in these cities were of four kinds, designated by Mr. Cushing as follows: 1, priest temples; (2) sun temples; (3) communal dwellings and (4) ultra-mural houses.

The priest temples.—These were the most conspicuous buildings in the ancient cities. As a rule there was only one to each city, and this was centrally located; but in one of the cities observed there were seven such buildings, the largest of which was centrally located. The reasons for this peculiar distribution, Mr. Cushing believes, are explained by Zuñi folklore and modern Zuñi customs. The ruins gave evidence that the buildings, when standing, were many stories high—from four to seven stories it is estimated. The Casa Grande on the Gila is said to show traces of five floors in that portion of its walls which still remain, and it is probable that one or two stories have fallen. Each building was surrounded by a high rectangular wall from 5 to 10 feet thick. A portion of this wall remains, and, being filled with the *débris* of the fallen building within, lends to the mound-like ruin that terraced appearance before alluded to. The lower story in each building was divided into six apartments, four great and two lesser. These apartments, the explorer believes, were used as store rooms for the priestly tithes in maize, etc. The other stories are supposed to have been used for priestly residences and for sacerdotal purposes. The entire building is thought to have served, not only as a storehouse and temple, but as a fortress in times of danger. Besides these in Arizona, there are great houses of similar construction in Sonora and Chihuahua, in northern Mexico.

The manner in which these buildings were constructed is perhaps peculiar. They might be regarded as great mud-covered baskets. For the thicker walls two rows of posts were erected and secured, one post to another, in different directions, by means of smaller sticks firmly lashed to them. The framework thus constructed was wattled with reeds, so as to form two upright hurdles braced together. The space between these was filled with well-packed mud, and the hurdles were thickly plastered within and without with the same substance. The thickness of the wall depended on the distance between the hurdles. For the thinnest walls, the internal partitions, but one hurdle was erected, and this was plastered on both sides. These structures of wood and reed no longer remained when the excavations were made, but the cavities found in the walls gave evidence of their former existence.

Sun temples.—The buildings which Mr. Cushing designates by this name, though not as lofty as the priest temples, covered a greater superficial area. The smallest measured was 50 feet in width by nearly 100 feet in length. One was discovered whose dimensions were about 150 feet in width by over 200 feet in length. Like the priest-temples they were built of earth on a great basket form or frame of hurdles; but the basket form instead of being rectangular was elliptical in shape. There is evidence that this frame of hurdles gradually tapered toward the top, and that the structure was roofed in with a dome made of a spirally contracting coil of reeds, resembling the coil baskets now so commonly made by the various tribes of the southwest. This spiral coil, as well as the rest of the frame, was heavily covered outside with mud, so that the structure when finished must have appeared, as Mr. Cushing expresses it, like an unburned, inverted and elongated terra-cotta bowl. The floor was elevated at its edges so as to form a sort of amphitheater and in the center was a hearth. It is thought that in these buildings the public rites of esoteric societies were performed as well as the sun drama and other ceremonies. The sun temples were usually in close proximity to the priest temples, and their ruins presented the appearance of low oval mounds depressed in the center.

Communal houses.—The great structures thus designated were the principal dwelling places. They were built of mud without the central frame of hurdles on which the walls of the temples were raised. They contained many rooms on the ground floor, and, as there is evidence that they were sometimes more than one story high, it is not improbable that they resembled much the modern terraced pueblos of New Mexico and Arizona. They were too large for the dwellings of single families, and for this and other reasons they are thought to have been each the home of a separate gens, clan, or some other large subtribal division. Each was surrounded by a separate high earthen wall and generally by a separate canal or acequia, although, in a few instances, two or more communal dwellings were included in the same encircling canal. Each had its single appropriate water reservoir with a branch canal leading into it, its one separate pyral mound or place of cremation, and its one great underground oven for the preparation of food. In Los Muertos at least fifty of these great buildings were wholly or partially unearthed, and it is likely that many more remained unrevealed beneath the surface of the ground.

Ultra-mural houses.—These were small, low huts, not rectangular in form, made of sticks, reeds, and similar perishable material, lightly coated with mud, and they probably resembled much the modern *jakal* or hut of the lower classes in many parts of Mexico, or the houses of the present Pima Indians of the Gila Valley. Mr. Cushing calls them ultra-mural or ultra-urban because they were situated outside the limits of the towns of earthen houses and not mingled with them; they formed separate groups. He conjectures that they may have been residences of an outcast population such as exists at Zuñi to-day. As each contained a central fireplace it is evident that they were occupied in winter as well as in summer, and were, therefore, not like certain houses scattered through the fields of the modern Zuñis, used only as temporary shelter for laborers while the crops are growing. These ultra-mural dwellings were very numerous; in one place constituting, of themselves, a town of considerable size, which contained a sun temple but no priest temple. In estimating the age and character of some, at least, of these houses, it must not be forgotten that as late as the seventeenth and eighteenth centuries we have records of the existence of Pima villages in the lower part of the Salt River Valley. I make this statement on the authority of Mr. Bandelier.

AGRICULTURE AND WATER SUPPLY.

When these ruins were inhabited cities, the land in which they lie was, as it now is, an arid region, where agriculture could not be conducted without irrigation. The works constructed by the ancient inhabitants to establish irrigation are as noteworthy monuments to their industry and intelligence as are their stupendous buildings. The explorers have traced in this particular realm in the Salado Valley, they estimate, over 150 miles of the larger canals—the mother acequias or *acequias madres*, as the Spanish-Americans call them. Their remains have been found at distances of 12 and 15 miles from the present bed of the river, and there is no evidence that the river has materially changed its course since the days of the ancient inhabitants. The miles of smaller acequias could not be estimated.

The larger canals varied in width from 10 to 30 feet and in depth from 3 to 12 feet. Their banks were terraced in such form as to secure always a uniform central current in the canal when the rains ceased in the mountains and the waters diminished. It is thought that this device was to facilitate navigation, and that the canals were used not only for irrigation, but for the transportation of the produce of the fields and of the great timbers from the mountains which the people must have needed in the construction of their tall temples and other houses.

In various parts of our arid region the old Indian canals may be still easily traced where they are cut through hard soil or where they are so exposed and situated, with regard to the prevailing winds, that the sand is blown out of them rather than drifted into them. There are places in Arizona where the American settlers utilize old canal beds for wagon roads. But in most cases the canals have been filled with sand and clay to the level of the surrounding soil and, to the ordinary observer, no vestige of them remains. Yet Mr. Cushing, guided by his knowledge of a custom which exists among the Zuñi Indians, was able to trace the course of these obliterated channels. These Indians, he relates, have observed that wherever there is running water there are rounded pebbles and boulders; reasoning, as man is so apt to do, inversely to the natural order of causation, they suppose not that the waters shape and deposit the pebbles, but that the pebbles control and direct the flow of the waters. For this reason they place such stones on the margins of their artificial water courses to hasten and direct the current. The presence of these pebbles disposed in lines, at the surface of the ground, caused the explorer to surmise that they marked the sites of irrigating ditches, and excavation proved the surmise to be correct. Pebbles which had once been used as implements and become worn out or broken in service were those most usually employed for this purpose.

Within the past twenty years, since the wild Indians of western Arizona have been subdued and order has been established within that region, the locality in which Los Muertos and its neighboring cities lay has been again restored to cultivation—this time by the white race, who utilize, through new channels, the waters of the same Salt River that fed the fields of the departed races. The canals of the moderns follow straight lines; those of the ancients were tortuous; but the ancient people used the water to greater advantage than their successors and covered with their system a wider territory. In the old canals the fall was about 1 foot to the mile, in the new it is 2 feet to the mile. The ancients constructed great reservoirs to store the excess of water when the river was high; the present occupants have no such works. Since this region has been reclaimed it has proved one of the most fruitful within the boundaries of the United States and is adapted to a wide range of vegetation, temperate and tropical.

In one place, near the present Mormon settlement of Mesa City, about 10 miles from the ruins of Los Muertos, the canal was dug through a hard, rocky layer. The Mormon community made use of the prehistoric cut when constructing their own irrigating ditch. I have heard on good authority that the Mormons estimate the labor thus saved to them at $20,000. Who will calculate the equivalent of this in human hands and days of work during the age of stone and when man was his own beast of burthen?

In addition to the river irrigation the ancient Saladoans had a system of rain-water irrigation. In the woodless mountains immediately surrounding their homes, the Superstition Mountains, the Estrellas Mountains, etc., brief but heavy rains sometimes fall, which flow at once into the plain, causing heavy floods and doing more damage than benefit to the crops. In these mountains there are neither springs nor constant water courses and only a desert flora. The heights which give birth to the Salado and the Gila are farther away and of much greater altitude. To conserve the waters of these sudden rains in the neighboring hills the people built dams in the ravines and large reservoirs in suitable places in and near the neighboring foothills. From these reservoirs the waters were, when needed, allowed to flow gradually over the fields. This may be regarded as evidence that the waters of the rivers, abundant though they were, were not sufficient for the needs of the population.

BURIALS.

The bodies of the dead were buried both with and without previous cremation. Those buried without cremation were always buried in the houses, either under the ground floors or in the walls. The cremated remains were interred outside of the houses.

The wall or mural burials were found mostly in the priest temples, in what remained of the first and second stories; a few were discovered in the communal dwellings. The body in such a burial was inclosed in an adobe case, and a niche was cut in the wall for its reception, which was afterwards filled and plastered over with mud, so as to leave no external evidence of the burial.

The burials under the floors were confined to the communal dwellings. The graves were constructed with different degrees of care; the more perfect being rectangular holes carefully plastered on the sides with mud and sealed over with the same material. The dead were usually placed with their heads to the east and slightly raised or pillowed so that the faces were turned toward the west. The hands were laid at the sides or over the breast. The lower extremities were placed as we place those of our dead except in one instance, that of an adolescent female who was supposed to have been sacrificed to the gods to avert earthquake. She was buried with the limbs abducted.

In a few instances in the communal dwellings the body was buried partly under the floor and partly in the wall. This was supposed to be for the purpose of economizing space. The trunk, in a supine position was buried close to the wall; the lower limbs, elevated at right angles to the trunk, were placed in a niche in the wall which was then filled up with mud.

Among those buried under the floors, many were children, and these were found always buried near the kitchen hearths. This is a custom which is found to have prevailed in other parts of the world and is variously accounted for. Mr. Cushing's explanation derived from Zuñi folklore and belief is this: "The matriarchal grandmother or matron of the household deities is the fire. It is considered the guardian as it is also, being used for cooking, the principal 'source of life' of the family. The little children, being considered unable to care for themselves, were placed, literally, under the protection of the family fire that their soul-life might be nourished, sustained, and increased."

FIG. 4.—Pyral cemetery, unearthed.

Within both the underground and wall sepulchers were found deposited various household utensils, articles of personal adornment and others of a sacerdotal character. In the mural burials of the temples the articles of sacerdotal use were particularly numerous and elaborate. This is one of the many reasons Mr. Cushing has for believing that those buried without cremation were of a sacerdotal and higher class of the community, while those who were cremated were of a lower class, and laymen. The pottery buried with the adults in the graves, was left whole and not broken or "killed" in the manner to be described when speaking of burials after cremation; that buried in graves with children was, however usually "killed" or broken. The sacred parapher nalia referred to were so similar to those used in Zuñi to-day that Mr. Cushing "was often able, through the knowledge of the Zuñi priesthoods to identify the medicine or priestly rank of the silent occupant of a sepulcher."

The great majority of the dead were cremated. Each communal dwelling had in close prox imity to it, its own pyral mound and, situated at the base of the latter, a collection of earthen vessels containing the remains of the dead—a pyral cemetery (Fig. 4). The mounds consisted of ashes, cinders, and fragments of charred and broken mortuary sacrifices; they were from 60 to 100 feet in diameter, from 3 to 9 feet high and showed evidence of having had from 2 to 6 locations for pyres in each. That each pyral mound was appropriate to its neighboring communal house was inferred from the correspondence of certain special marks and designs on the pottery in the pyral cemetery with designs found on pottery in the graves of the contiguous dwellings.

The burnt bones and charred remains of some of the more valued articles of personal property were placed in pots of suitable size, which were covered by inverted bowls or broken pieces of pottery and surrounded by other articles of pottery buried as presents to the dead. These mortuary gifts were broken or drilled before burial, probably in order that the souls they were thought to possess might escape and accompany the dead to the spirit land. The custom of breaking the pottery sacrificed with the dead is called by the people of modern Zuñi "killing" the vessels, and is still practiced among them.

It is believed that those of the priestly race were not cremated because they had the power to release their own souls from their bodies while the laity, having no such power, had to have their bodies burned to effect the desired release. Whatever may have been the creed that thus preserved some bodies for simple interment, anthropology owes it gratitude, for without it the unique skeletons of this archaic race would not have been preserved for modern study and comparison. It is thought, too, that the pots buried with the uncremated adults were not broken or "killed" because the priests knew how to release the souls of the pots and take them with them to the undiscovered country, while to the laity such knowledge was denied.

Fig. 5.—Double burial.

Double burials were found both with the cremated and the uncremated remains; but were much more common with the latter than with the former. When two skeletons were discovered in one grave or incinerary vessel they were invariably adult, and, whenever the sex could be determined, one was always found to be a male and the other a female—presumably man and wife. This might be thought to indicate that the wife had been sacrificed at the death of the husband; but in the house-graves there was often evidence that the interments were not simultaneous, the upper grave not being dug exactly over the lower and the bodies having been apparently wrapped in different cerements. It was a rare thing to find three buried in one grave. Fig. 5 shows a double burial, male and female, in which the interments, and probably the deaths, were simultaneous.

ARTS.

Nearly all the implements and tools discovered were of stone, but of beautiful finish and great variety of form. No metal tools, whatever, were found. The only articles of metal were little rude copper bells.

FIG. 6.—Small water-jar, found in fourth sepulcher, buried with child, in Los Muertos.

A copper bell consisted of a plate of the metal wrought into leaflets. These leaflets were brought together at the apices so as to form a hollow ball with meridional openings. In this ball a pebble was imprisoned for a clapper. The handle, or stem, was soldered on in a manner which

FIG. 7.—Small water-jar, found buried with child in house sepulcher, southern portion of Halonawan, ancient Cibola.

indicated a knowledge of a soldering material and the use of the blowpipe; and indications are not wanting that the bells were not introduced from a distance by trade, but were manufactured where found.

Pottery was found in great abundance in the house graves, in the pyral cemeteries, and on the floors of the houses, where it seemed to have been abandoned, as if the dwellings were suddenly deserted. It consisted of food vessels and water vessels in a great variety of shapes and sizes, and of well-executed images of animals of the chase which once inhabited the surrounding

FIG. 8.—Ancient Cibola eating bowl, showing "exit trail of life."

country. The **vessels were** decorated **in a** manner closely resembling those of the modern **Pueblos** of New Mexico and Arizona, especially those of the Zuñi and Moqui (see Figs. 6, 7, 8, 9, and 10). The more commonly employed symbolic decorations were alike in all.

FIG. 9.—Modern Zuñi food bowl, showing "exit trail of life."

FIG. 10.—Modern Zuñi water vessel, showing "exit trail of life."

One of these, worthy of especial note, is what the Zuñis call the exit trail of life. It is found inside of **food vessels and outside** of water vessels; it consists of an opening or hiatus in the single or double **encircling paint bands near** the margin of the vessel, as shown at *a* in Figs. 8, 9, 10. It is a symbol based **on the idea before alluded** to of vessels having souls.*

* See Fourth Annual Report of the Bureau of Ethnology, p. 510.

Another decoration, shown in Figs. 11, 12, 13, may here be mentioned. It is undoubtedly an animal figure which in textiles and basketry has been necessarily conventionalized into a figure

Fig. 11.—Ornamental zone on water jar from Los Muertos.

bounded by straight lines, and from the woven forms transferred, more or less modified, in paint to the pottery. It is common on both the ancient and modern pottery of our southwestern land of ruins, and is frequently seen in the cloths of ancient Peru. (See Fig. 14.)

Fig. 12.—Symbolic decoration in white-bordered black, adapted from ornamental zone on water jar of red slip ware from Los Muertos.

The articles of personal adornment which remain are principally of shell and consist of rings, bracelets, pendants, etc. Some of these were ornamented with geometrical designs and inlaid

Fig. 13.—Medium sized eating bowl of red slip ware, with white-bordered black paint decoration. From Halonawan, one of the ancient seven cities of Cibola.

with turquois and other precious or semiprecious stones. Sea shells carved in the form of a frog were common and one or two of these frog images were beautifully inlaid with turquois and other stones of brilliant color. The inlaying was accomplished by coating the shell with some black vegetable gum (supposed to be that of greasewood) which hardened on drying; the gems were stuck into this coating and, when the latter became hard, the whole was rubbed down to a smooth surface. An accurate chromo-lithographic illustration of one of these artistic objects has appeared in *Gems and Precious Stones of North America*, by George Frederick Kunz, New York, 1890.

Everything susceptible of decay in these ruins had disappeared; hence, with two or three trifling exceptions of charred and defaced articles, nothing was left of their woven stuffs, their basketry, their woodwork, or the featherwork. But that they wove cloth, wrought baskets and made useful and ornamental objects in various perishable materials, we have abundance of collateral evidence.

During the first fifteen months of the work of the expedition from 17,000 to 20,000 specimens of various kinds were collected, and many fragments rejected. But the collection would have been far richer were it not for the wanton destruction of much material by visitors. Sometimes when

Fig. 14.—Mummy from cemetery at Ancon, Peru.

a pyral cemetery or the floor of a large dwelling had been unearthed, and all the articles discovered laid in their original positions to be photographed, a party of sightseers would appear and, either in the absence of the workmen or in spite of their remonstrances when present, trample the objects under foot or deliberately kick the pottery to pieces to "see what was inside." In the earlier days of the work many fine skeletons were lost in this way. Some persons even appropriated handsome objects and carried them away, maintaining that, as these things were found on public land, all had an equal right to them.

POPULATION.

What was the population which in ancient days subsisted on the crops watered by the Salado or Salt River and the stored rains of the neighboring mountains? What was the population of the old Salado settlement? Opinion is divided on this subject, and will probably long continue to be divided. Some who have had the best opportunities of observing the ancient works and studying the problem estimate the population at from 80,000 to 100,000 souls. Los Muertos, it is calculated, covered an area of over 2 square miles and contained about 13,000 inhabitants. There were six other groups of buildings in the region as large or larger than this, and there are indications that they were simultaneously occupied. If it could be shown that they were not occupied at the same time, a much lower estimate of the population would have to be made. As the land is now becoming rapidly filled with white settlers, and the ancient town sites are being covered with farms and crossed with irrigating ditches, all antiquarian problems become more difficult of solution every day.

ANTIQUITY.

In 1539, when Friar Marcos made his journey to Zuñi, and when, a year later, Coronado marched with an army to the same point, they passed within about 100 miles of these towns. Had they been inhabited in those days, the travelers would doubtless have heard of them, for the fame of the less significant Seven Cities of Cibola reached them in the heart of Mexico and induced them to travel 200 miles further northeast than the mouth of the Salado. They were ruins, no doubt, three hundred and fifty years ago, or at the beginning of the historic period of Arizona. No vestige of anything belonging to the iron age or of European origin was brought to light in the excavation. The writer knows of other ruins in New Mexico and Arizona which, from recorded

FIG. 15.—Skeleton of man supposed to have been killed by earthquake.

evidence, are known to have fallen to decay and been abandoned long before the historic period; yet in these textile fabrics and other perishable articles are still found fairly preserved, and particularly the hair of the dead has survived the process of decay. In Los Muertos were found no hair, no cerements, nothing that might have escaped destruction in a thousand years. It is thought by Mr. Cushing that from one thousand to two thousand years may easily have elapsed since the priests of Los Muertos worshiped in its standing temples. The Casa Grande of the Gila was a ruin standing in the sixteenth century probably much as it stands to-day; three and a half centuries have wrought little change in it; but the similar priest-temples of the neighboring Salt River are mere mounds of earth. The writer has seen two photographs of the Casa Grande of the Gila taken from the same point of view, one twenty years after the other; yet in the pictures no difference can be discerned in the most minute points and prominences of the ruin, which were subject only to the modifying influences of rain and wind, though the parts within the easy reach of human hands have suffered notably.

It must be remembered that earthquake may have hastened the fall of the Salado temples. The explorers have found many indications that these cities were abandoned on account of earthquake, and Zuñi myth and tradition point to former migrations of the people induced by seismic disturbances. One skeleton in Los Muertos was found lying on its face, evidently of a person never formally buried, and apparently crushed by falling walls.* (See Fig. 15.)

Fig. 16.—Outline drawing, full size, of terra cotta image of animal, supposed to be allied to the vicuña.

It has been indicated in the previous pages that an intimate relationship in arts, civilization, religion, etc., has been found to exist between the ancient Saladoans and the ancient sedentary people of Arizona and New Mexico in general, as well as the still extant sedentary tribes of this region. A relationship, less intimate perhaps, may be shown to exist between them and the

Fig. 17.—Rock inscription thought to represent vicuña-like animals and man throwing bolas.

ancient house-building tribes of old Mexico and Central America. There are many facts, too, which point to a close connection between the Saladoans and the ancient Peruvians—a connection more close perhaps than that between the former and many races who lived nearer to them, geographically, than the Peruvians. Environment may have had its influence on this affinity, for

* Since the above was written it has become apparent that we may attribute the sudden destruction of these earthen buildings to floods as reasonably as to earthquakes. In the spring of 1891 this region was visited by a great flood, which covered much of the Salt River flood-plain and ruined many of the adobe houses of the white settlers.

the conditions surrounding human life in Peru are more like those of Arizona than those of tropical Mexico and Central America. The following are some of the indications of a special relationship between the ancient Peruvians and the ancient Arizonians:

1. In the ruins of some of the ultra-mural houses there were unearthed terra-cotta images of a quadruped which can not be identified as resembling any animal of the present North American fauna, while all other effigies found are easily identified. Unfortunately I am able to present only an outline drawing of one of these (Fig. 16.) Zoologists who have seen the original terra-cottas are of the opinion that it represents a creature allied to the South American *camellidæ* (llama vicuña, guanaco, etc.). In various parts of the Southwest there are petrographs which are thought to represent the same animal. Some of these petrographs are located at considerable distances from Los Muertos, as, for instance, those in the Puerco Valley, some 250 miles away.

FIG. 14.—Rock inscription representing, it is supposed, vicuña-like animals and bola-throwers; besides deer and other animals.

It has been surmised that such animals continued to be domesticated by the sedentary Indians of Arizona and New Mexico down to historic days and became extinct only when the more serviceable European sheep was introduced by the Spaniards. This surmise is based on certain statements found in the works of early writers and explorers who speak of the Pueblo Indians having a coarse cloth, something like woolen cloth, and having small wool-bearing animals domesticated in their houses. But Prof. Bandelier, who has studied the early documentary evidence relating to the Southwest more thoroughly, no doubt, than any other living student, discredits the modern existence of these animals. In a letter to the writer he shows that we have only hearsay testimony as to their existence and concludes with these words: "If there has ever been a llama, guanaco, or vicuña, known to the Southwestern Indians, it became extinct long previous to the sixteenth century." Fossil bones of an animal of this family have been found in the Southwest; but its bones were not identified in the Salado ruins.

2. In several places among the ruins, on the floors of the houses, near the walls (as if they had fallen from the latter), were seen peculiar groups of stones, consisting of three globoid and one ovoid pebble. These are thought to have been the stones of *bolas* such as are now used in South America to catch wild or half-domesticated animals. The buckskin cases and thongs which connected the stones are supposed to have decayed, like all similar material in the ruins. The presence of these stones would, in itself, be insufficient evidence of the use of *bolas* among this people, but

it is thought that **the** petrographs afford additional testimony. Where these vicuña-like animals **are** delineated on the ancient rock-carvings, they are often associated with the figure **of** a man holding in his hand a peculiar four-branched instrument; one of the branches is held by its extremity **in** the hand, the others are in the air (Figs. 17, 18, and 19). This is thought to depict a herdsman 'or hunter in the act of casting the *bolas*. The *bolas* have, as far as can be ascertained, not been in use in North America south of the Artic circle since the Columbian discovery, although **an** implement, analogous in use but different in form, is employed by the Eskimo.

FIG. 19.—Rock inscription of supposed bola-thrower, dancing men, and other objects.

Fig. 20 is a copy of a rock inscription showing a number of these animals associated with a hunter bearing a bow. Fig. 21, also from a rock carving, represents a supposed bola-thrower in connection with a flock of turkeys. The turkey is found wild in Arizona and was probably **domes-**ticated by the ancient inhabitants of the country.

FIG. 20.—Rock inscription of vicuña-like animals and hunter.

3. In sacrificial caves in mountains surrounding the Salado cities, knotted cords **have been** found which are much like the *quipus* used by the ancient Peruvians. Similar knotted cords are still in use by the people of Zuñi and are by them called kispuwe, a name very **similar in** sound to the Peruvian.

4. In addition to these indications we must consider the great and closely corresponding prevalence of the *os Incæ* in the skulls of these two widely separated peoples. This is a subject discussed more fully in the somatological part of this work.

EXPLORATIONS NEAR ZUÑI.

The expedition left the Salt River Valley in June, 1888, and arrived at Zuñi in the following month. The work was continued in the vicinity of this place under the direction of Mr. Cushing until October 20, 1888, when he left for the East. His physical condition was such that he was not able to return; but the work of excavation was continued in his absence until July, 1889, at which time the expedition was disbanded.

The location of the Seven Cities of Cibola, visited by Coronado in 1541, was long a question for scientific discussion, and many arguments were advanced in favor of different places; but the ethnologic researches of Mr. Cushing and the historical investigations of Prof. Ad. F. Bandelier have settled the question beyond a reasonable doubt. The Seven Cities were situated in the valley of the Zuñi River in the neighborhood of the present pueblo of Zuñi, in Valencia County, N. Mex. The accompanying map (Fig. 22) shows very approximately the location of each. It was prepared with the assistance of Mr. F. Webb Hodge, of Washington, formerly secretary of the expedition.

FIG. 21.—Rock inscription, turkeys, supposed hola-thrower, etc.

We give below a list of the names of the cities in the modern Zuñi language, as noted by Mr. Cushing, and in the old Zuñi or Cibola language, as noted by Coronado and other Spanish travelers and writers of the sixteenth and seventeenth centuries. If we make due allowance for the difference between a carefully devised modern orthography and a haphazard spelling of three hundred years ago we need not suppose that the language of Cibola has changed materially during the intervening time.

NAMES OF THE SEVEN CITIES OF CIBOLA.

Modern.	Sixteenth and seventeenth centuries.
1. Hawiku	Ahacus, Avicn, Aquico, Jahuicu, Havico.
2. Kyánawa, Hampasawan	Cunabe.
3. Kètchupawc, Kwákina	
4. Apina, Pinawan	Aquinsa.
5. Halona, Hálonawan	Alona.
6. Mátsaki	Mazaqui, Maçaque, Maçaquia.
7. Kyákina	Caquina, Kyakina.

It was the original intention to explore all the ruins of the Seven Cities; but the illness of the director and the consequent recall of the expedition prevented the fulfillment of this plan. Only one of the ruins of the seven cities was explored to any extent, namely, the ruin of Halona. This town occupied in part at least the site of the present pueblo of Zuñi. The excavations were made

upon the opposite bank, from Zuñi, of that meager and inconstant desert streamlet known as the
Zuñi River and in the neighborhood of houses occupied by the present ultra-urban population of
the Zuñi tribe.

Explorations were conducted in other ruins in the neighborhood. Some slight digging was
done in those on the top of Inscription Rock; but the most work was accomplished at Héshota-
úthla, a ruin on the road to Wingate, some 12 miles in a northwesterly direction from Zuñi.
Héshota-úthla was in its day a compactly built, many-storied stronghold of stone containing a

Fig. 22.—Zuñi towns, ruins of Cibola and other ruins.

population of probably more than a thousand people. It was not one of the Seven Cities; but,
according to the traditions (corroborated by archæological investigation) of the Zuñi Indians, it
was occupied by their people in a remote antiquity. From this ruin was derived the greater part
of the "Cibola" skeletons described in the second part of the following report.

In preparing this introduction, the writer has had access to some of Mr. Cushing's notes,
especially to the original manuscript of a paper contributed to the Berlin meeting of the Congress
of Americanists in October, 1888, and he has consulted a pamphlet entitled "The old New World,"
an account of the explorations of the Hemenway Southwestern Archæological Expedition in

1887-'88, by Sylvester Baxter (Salem, 1888). In addition to all this he has had the advantage of a personal knowledge of the southwestern country, its antiquities and its people, extending over a period of ten years. He has had an equally long intimate personal acquaintance with the director of the Hemenway Expedition. In the autumn of 1887 he had the rare good fortune to spend about a month with Mr. Cushing at Camp Hemenway, in the Salado Valley, while the excavations at Los Muertos were being carried on.

He **might**, therefore, had he so desired, have made of this introduction a more extensive and pretentious essay. This is intended, however, not as a contribution to American archæology, but merely **for** the convenience of the anthropologist who may desire to know something of the people to **the** description of whose osseous remains this work is chiefly devoted. The author has introduced only some of the more easily explained discoveries of the expedition, and he has made many statements without setting forth all the facts and arguments on which they are based. The reader must take some things for granted until Mr. Cushing's final report appears. In referring to the early Spanish writers and travelers the writer has been obliged to omit the proper bibliographical notes, for the reason that he had not access to their books at the time of writing.

In studying the crania and other bones described in the following pages, and in preparing this report, I must acknowledge my great indebtedness to the following gentlemen of the staff of the Army Medical Museum: To Dr. Jacob L. Wortman (who spent many months in the field collecting and preserving the bones), for assistance in preparing the sections on the teeth and hyoid bone; to Dr. D. S. Lamb, for assistance in preparing the section on the olecranon perforation; to Mr. Porter Tracy, for his labor in taking measurements and his help in many other ways, and to Dr. J. C. McConnell, for preparing the illustrations.

WASHINGTON MATTHEWS,
Surgeon, U. S. Army.

FORT WINGATE, NEW MEXICO,
September 1, 1890.

S. Mis. 169——11

HUMAN BONES

OF

THE HEMENWAY COLLECTION.

PART I.

THE SERIES OF SALADO.

HUMAN BONES OF THE HEMENWAY COLLECTION.

PART I.—THE SERIES OF SALADO.

§ 1. CONDITION AND REPAIR OF BONES.

As we have stated in the introduction, the bones when found were in an advanced state of decay and exceedingly fragile; particularly was this the case at Los Muertos. The organic remains at Los Hornos, Los Guanacos, and Las Acequias were usually in better condition than at the first-named ruins, owing, probably, to the greater dryness or other more advantageous quality in the soil. At Las Acequias they were in the best condition of all. When carefully unearthed the bones, *in situ*, in the graves might seem in sound condition, but the slightest manipulation—a touch of the finger even—would cause them to crumble into dust. The bones of the upper face, the pelvis, and the epiphyses of the long bones were the most friable. Parts successfully unearthed, but not immediately conserved, if they escaped the despoiling foot of the mischievous visitor, would often disintegrate in a day or two from the effects of exposure to sun and wind. After a period of annoying experiences it became the custom to apply paraffin, shellac, or other preserving substance to the bones before their removal from the graves, or immediately after.

The skulls were nearly all obtained in a fragmentary condition; the fragments, carefully packed, were forwarded to the Army Medical Museum in Washington, and here a number were, with much labor, put together in such a manner that they might be measured and studied as entire skulls. The remaining fragments often gave us valuable points for anatomical study. In the work of restoration we had in many cases to use plaster of Paris to fill gaps or strengthen weak parts. Where the plaster was used superficially to replace thin scales from the outer table, measurements were, after due deliberation, sometimes taken from points on the plastered surface; but where the plaster had been thickly applied, had been used in restoring salient points, or had been employed to fill a gap in both tables of the bone, its presence was considered to preclude measurement. In a small number of skulls where we had, after restoration, reason to suspect the existence of post-mortem distortion, measurements were not made—not, at least, in the regions affected by the distortion. A great but unavoidable disadvantage in the use of the plaster was that it encroached on the cranial cavity and thus usually rendered the cubature of the latter impossible.

§ 2. THE MEASUREMENTS OF THE SKULLS.

In preparing this report we have kept two objects in view: First, that we might obtain material for our own study and comparison of this collection; and, second, that we might furnish to other investigators material for comparative study. In providing for the latter we have taken some measurements which we have not used as data for subsequent investigations, and we have not confined ourselves to the methods of any particular school or system.

For purposes of our own research we have employed chiefly the measurements of the French and English schools of anthropology as formulated in Dr. Paul Topinard's recent work,* because the literature of anthropology is richest in studies based on these measurements (Appendix A), and the opportunities for comparison with them is consequently most extensive.

Recognizing the fact that a great number of anthropologists throughout the world have signified their intention of employing the measurements proposed by certain German anthropologists, formulated in what is known as the Frankfort agreement (Appendix B), and hence,

---

*Topinard; Éléments d'Anthropologie Générale; Paris, 1885; chap. XXVII.

anticipating an extended use of this **system in** the future, we have taken many measurements according to this agreement.

Our decisions **as to adopt and what to** discard in different systems may **appear** occasionally somewhat arbitrary; **but they** have usually been made in accordance with certain rules which we have been constrained to adopt. We have not undertaken to sit in judgment **on** the general relative merits of any system. All the systems extant are the results of more thought and study than we have been able to devote to the subject of craniometry. We have accepted that which seemed best suited to the scope of our work and to the character and extent of the series to be studied. We have also had to take into consideration the limited time at our disposal.

Any measurement which we believed to be identical or practically the same in different systems we have taken but once and, in taking it, we have followed whichever rule seemed the most explicit or laid down the most definite landmarks. Thus, in taking such a short dimension as the interorbital width, where a small error may count for much, we have chosen for our landmark the definite dacryon as directed by Broca, instead of the less certain "inner border of the orbit," which the Frankfort agreement prescribes for us.

On some occasions we have discarded a dimension which had been made, or might be made, the basis of extensive and valuable study, for the simple reason that we did not regard the given directions as sufficiently explicit. While they might be clear to the scientist who wrote them, or to the student who saw him apply them, they were not clear to the reader who had nothing but the text to guide him. Thus we have taken no vertical measurements from the ophryon, because **no** one tells us in what plane the connecting line between the frontal ridges should lie. Two or more equally short lines between these ridges might, in some skulls, be described at some distance from one another on the median line. In other words, we might have two ophryons so far apart as to give materially different facial heights. We thought it better to be silent than to fill our pages with material which might prove worthless. Had we had unlimited space, time, and assist**ance we** might have included such measurements, though not without comment.

Some measurements were forbidden by the character of the skulls. The histephanic diameter was put aside because the temporal ridges in this series are indistinct, the indistinctness being due, possibly, to the general use of boiled vegetable food among the Saladoans and the consequent limited exercise of the temporal muscles. Moreover, the stephanion lies in a region especially liable to be broken, and frequently was broken in the series. We have substituted for this dimension the maximum frontal diameter of Emil Schmidt.[*] The upper incisors were so often missing that we neglected dimensions into which they entered. On the other hand, we took measurements from the metopion, which is a very uncertain guiding point on these skulls in consequence **of** the subdued character of the frontal bosses.

We felt a great temptation to present to the reader such **opinions concerning all** the meas**urements** as we formed in the progress of our work, **and to give our reasons in** each case for **adopting** this or abandoning that method; but on mature reflection **we felt that** this would lead **us beyond** the proper scope of our work. In the more important **cases,** comments on the methods **are** given in connection with the discussions of particular dimensions or indices. In some instances **we** advanced far in the **work** of securing a dimension **before we** found practical reasons for abandoning it. In other **cases** we have taken a measurement **on all** the skulls of the series and com-piled our tables **and** indices before we concluded to suppress our results. This we did, for instance, in the case of the horizontal and vertical measurements of the orbit.

We have, with some inconsistency, perhaps, adopted dimensions and followed rules of **whose** exactness we felt no less uncertain than we felt of some which we discarded. Such instances are, perhaps, to be classed among our arbitrary decisions. But we can partly atone for our errors, if such exist, by telling exactly what our own methods were. For instance, we have recorded meas-urements which have the superior border of the meatus auditorius for a guiding point, and we must confess that we know not where to locate this point with accuracy. The rule for our own

[*] EMIL SCHMIDT: Catalog der im anatomischen Institut der Universität Leipzig aufgestellten craniologischen Sammlung; Archiv für Anthropologie, Braunschweig, 1887-'88, p. v.

guidance has been the contour of the processus auditorius or tympanic bone. Where this was complete in its upper portion, as it rarely is in man even in the lower races, we had no trouble in establishing our point. Where a good vestige of the upper part remained, not too far out of place, we were contented to take such vestige for our guide; but where a large segment of the bone was completely missing we joined the upper horns of the remaining portion by means of a pencil mark described as directly as possible from one horn to the other on the roof of the meatus and took the highest point of this arched line for our landmark. In leveling the skull for the German horizontal plane and in taking the auricular heights we felt less hesitancy in depending on this guiding point than in taking vertical arcs. Here it was most doubtful.

Table I is intended not only to answer the purposes of the present investigation, but possibly to serve as a model for future catalogues which may be issued by the Surgeon-General's Office. It has been designed with a view of economizing space and making reference easy. The peculiarities of its plan require little explanation. On the first page of the table and on its duplicate fly-leaf we have given a condensed description or indication of each measurement, index, or other item sufficient, we believe, for ready reference. In order to get each description within the space of one line we have rarely used the exact words of the original authors. For the measurement of the German anthropologists we have been especially compelled to reject the circumlocutions of the Frankfort agreement in describing guiding marks and have adopted instead the specific terminology of craniometric science. "Frankfort," in the table, denotes that the preceding rule is to be found, in substance, in the Frankfort agreement.* "Topinard" denotes that it is to be found in the work of this author already referred to. The number or letter which follows either of these names corresponds with that given to the item by the quoted authority; thus "Frankfort 1" refers to the first measurement of the Frankfort agreement. Feeling that our brief references to rules might often be insufficient for those who had not at hand copies of the oft-quoted Frankfort agreement and of the rules of Topinard, we have supplied these in Appendices A and B of this work.

§ 3. THE PICTURES OF THE SKULLS.

The outline tracings of the skulls shown in plates 1 to 54, inclusive, are reductions to half size, made by means of a pantograph from orthogonal or geometric drawings.

It seems proper that we should here describe the apparatus and the method in use for the past five years in the Army Medical Museum,* by which these orthogonal tracings were made, since both seem to differ in many respects from those in use elsewhere, as far as we may judge from published descriptions.

Fig. 23 represents the complete apparatus in use. It consists of a frame (a, a, a), inside of which is an open box (b) nearly filled with dry sharp sand (so arranged that it may be raised and lowered by means of a lever (c), a movable and adjustable mounted pin (d), an ordinary carpenter's or draughtsman's square, and a tracer of peculiar construction, which has been named the periglyph (e). The frame is surmounted by a movable plate of glass, thinly varnished on both sides to receive the tracing.

The periglyph is shown reduced in Fig. 24. It consists of a standard (a), a base (b) (both made preferably of vulcanite or hard wood), supported by two padded points (c), and by the sharp steel style (d), which makes the tracing; vertically above the extreme point of the style is a pin hole on an adjustable arm (e).

In other laboratories they use diopters, somewhat similar to this instrument in appearance, with which the outline is drawn by means of a pen or pencil held in hand. It needs but a single trial to convince one that our instrument, with its fixed steel tracer, is vastly more reliable and convenient. Of course the steel point would not trace on plain glass as the pen does; the thin coat of varnish renders the use of the style practicable.

---

* Verständigung über ein gemeinsames craniometrisches Verfahren; Archiv für Anthropologie, Bd. XV, Braunschweig, 1884, pp. 1–8. (See Appendix B.)

† W. MATTHEWS: Apparatus for tracing orthogonal projections of the skull, in the United States Army Medical Museum. Journal of Anatomy and Physiology, vol. XXI, London, 1886–'87, pp.42–45.

To ascertain if the pin hole is truly vertical to the apex of the style, we take sight through the former over the latter on some point of the object to be depicted under the glass, and wheel the instrument around—the point acting as the center—180 degrees. If the pin hole is vertical, the apex of the style will still cover the point on the object; if it is not vertical, we loosen the binding screw (f, Fig. 24) and adjust the arm (e).

The frame of our apparatus is 35^{cm} long, 28^{cm} broad, and 43^{cm} high; but some approximate size will do as well. The cross pieces which secure the upright supports are not placed at the top of the frame, but some 12^{cm} from it. There are two reasons for this: first, that no shadow may fall on the skull to obscure the vision of the operator, and second, that a horizontal surface may be afforded to support the mounted needle. The plane of the cross pieces must be perfectly parallel with that of the plate of glass.

FIG. 23.—Apparatus for orthogonal tracings.   FIG. 24.—The periglyph.

The mounted needle (Fig. 23, d) alone is used when the datum plane lies horizontally, as in outlining the vertex, side, and base of the skull; but when the plane stands vertically, as in tracing the anterior and posterior views, the square is also employed to secure the desired adjustment.

For facilitating the adjustment of the skull accurately and readily in any position, and for maintaining it in position, we have found nothing to excel the sand box. The most elaborate mechanical contrivance could not, we imagine, answer the purpose better.

In this series, furthermore, the skulls were so fragile that they did not admit of the application of any craniphore that would produce the least pressure.

The following is the method of operation: Place the skull on the bed of sand, pressing it down until it stands firmly. By means of the lever raise the sand box until the skull is nearly or quite on a level with the slots in which the glass is to fit. Orient the skull in the sand with the aid of the mounted needle, or the square, as the case may require; put the varnished glass in place; by means of the periglyph make the desired tracing; take off the glass. If a positive picture is desired, trace over the scratched drawing on the reversed side of the glass with ink. When the ink is dry, proceed to make the imprint. Lay unglazed paper on the inked figure and press it firmly down with one hand to prevent slipping; raise a small portion of the paper with the other hand; breathe in one spot upon the ink sufficiently to moisten it; replace the paper and rub it briskly over the moistened surface with the thumb nail. Treat the entire figure in this way. If a reverse picture is wanted, which is usually the case when a finished drawing is to be prepared for engraving, ink the scratched line and take the imprint therefrom.

If it is desired to prove the correctness of a positive picture, wash away the ink from which the imprint has been made, ink the scratched drawing and place it right side up over the positive on the paper. The two should correspond. In no instance where we have made this test have we found the slightest error.

We have used an ordinary black ink, and have been able to take three good impressions from one drawing. If it were desirable to take a large number of copies, other inks could be found to accomplish the purpose.

Dr. Paul Topinard tell us[*] that with Broca's stereograph the five views of the skull may be made in an hour. It takes nearly twice that time to do the same with our contrivance, operating with proper care; but as a partial compensation for this we have a drawing which furnishes many duplicates.

No special skill or lightness of hand is required with our apparatus; any person possessed of ordinary intelligence and eyesight can use it successfully at the first trial. It is not complicated; it requires no highly skilled workman to construct it; it may be made by any carpenter and its cost is insignificant.

Even the periglyph may be made by any handy individual with an ordinary pocket knife. We have two periglyphs, one manufactured of metal by a practical model-maker, the other rudely whittled out of wood by a medical gentleman connected with the Museum; both are perfectly accurate, but the latter is the more easily handled and the favorite instrument.

Several outlines may, without confusion, be drawn on the same varnished surface. The varnish should be of such a character that when dry it becomes crisp and brittle, breaking up in the course of the stylus—not dragging after the instrument and clogging it. Of many mixtures tried that known in the trade as Berry Brothers' (Detroit) hard-oil finish, diluted with one-third turpentine, gave the best results.

In making all but six of these tracings we adjusted the skulls on the German horizontal plane, or plane of the Frankfort agreement, partly for the reason that with the sand box we could find this plane more readily than we could find the alveolo-condylean plane. But for purposes of comparison we sketched the *norma verticalis* parallel with the alveolo-condylean plane in six specimens, the type skull and five which approximated the type. The reduced tracings are shown in plate 51.

The five views of the type skull[†] (Pls. 55–59, incl.) were made on the basis of elaborate orthogonal tracings, the shading being added by the artist from nature. They are natural size. It is greatly to be regretted that the nasal bones in the type skull were broken, and that we were obliged to make a plaster restoration. The shape of the nasal aperture is only approximate.

There were but few skulls in this series in which all the points of the German horizontal plane or any other horizontal plane could be found to coincide with a true horizon, while the sagittal plane was perfectly vertical to such horizon. The variance was most marked at the upper borders of the auditory meatuses. In order to approximate uniformity we always aligned our facial guiding marks, not with the upper margin of the right meatus, but with that of the left, the side on which the *norma lateralis* was taken.

The views of the lower jaws in plates 52, 53, and 54 were taken with the same apparatus and by the same method as were those of the skulls, and similarly reduced by the pantograph. When each was drawn the plate of glass on which the tracing was made was parallel to the plane on which the lower margin of the jaw rested at equilibrium.

Areas in the drawings marked with parallel straight lines show where there are holes in the skulls, neither bone nor plaster being present. Dotted areas indicate plaster restorations—all such repairs, whether deep or superficial, being thus shown.

---

[*] Op. cit., p. 253.　　　[†] For description see § 14.

§ 4. SEX.

In twenty-one cases the skulls are accompanied by enough of the remaining bones to let the sex be stated with considerable confidence. These twenty-one skulls are the following:

Males: Nos. H. 6, H. 7, H. 14, H. 18, H. 19, H. 24, H. 25, H. 26, H. 32, H. 34, H. 41—total, 11.

Females: H. 1, H. 5, H. 8, H. 10, H. 15, H. 21, H. 36, H. 39, H. 45, H. 57—total, 10.

These groups appear to be so scattered through the various ordinations that it can not with safety be said that the sexes are distinguished from each other by any metrical characteristics.

Although it is universally attempted to distinguish the sex of, say, four skulls in five, we do not consider it possible, in the present case at least, to do so; for, firstly, the number of known sex is so small that it is not possible to say that there is a constant sexual difference in any particular dimensional relation; secondly, there is apparently no constant difference of anatomical detail, such as prominent processes, "strong marking," or the like. The sex of H. 40, the type skull, can not be certainly stated, but, very reservedly, of course, we may suggest the probability of its being female. In this connection it is interesting to note that H. 7 and H. 25, males, also closely represent the type and closely resemble one another.

§ 5. PATHOLOGY.

Of the Salado collection about 69 sets of bones, representing each a complete individual skeleton, or the majority of bones of one individual, have come to us; but, as these sets are sometimes mixed with bones which do not belong to them, and as there are many miscellaneous bones in the collection, percentages of pathological formations must in some cases be only approximate. The collection shows some interesting anomalies, diseases, and injuries.

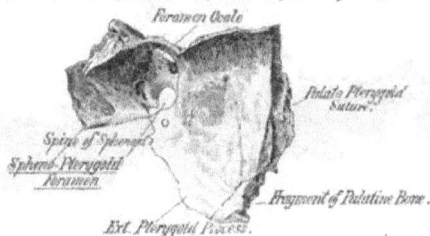

FIG. 25.—Fragment of skull, showing sphero-pterygoid foramen.

Anomalies.—The more important anomalies, those supposed to be of anthropological significance, are discussed more fully elsewhere, under separate headings. Some of those of minor importance will be considered here. In one case (fragment) the occipital bone showed two small, smooth, rounded condyloid prominences close to each other at the anterior part of the foramen magnum. The condyles proper were somewhat broken, but appeared to be smaller than usual, though normally located.

In one case (H. 21, Pl. 21) the foramen magnum was of unusual size. At least the portion of its border, the posterior half, which remained, indicated that the foramen was very large. The basilar portion of the occipital bone was missing.

The spheno-pterygoid foramen, complete or incomplete, was not found in any of the restored skulls; but in one small fragment a complete foramen was found, where the two processes which formed its boundaries touched but were not coössified (Fig. 25).

In H. 33 the occipital bone showed behind the right condyle, from which it was separated by a narrow groove, a small, smooth surface which articulated with a corresponding small, smooth surface on the atlas, behind the usual kidney-shaped articular surface. In the case of the atlas also there was an absence of the spinous process and of a small part of the posterior arch on each side, leaving a gap in the bone. There was no sign of inflammatory action.

In several instances the groove on the atlas for the suboccipital artery and nerve was converted into a foramen more or less complete, sometimes on both sides in the same subject. In H. 25 there are very complete foramina on both sides. In a number of cases the vertebral foramen was subdivided into two openings, and sometimes it was much smaller on one side than on the other.

Among the vertebræ there were five instances of what might have been congenital union; in two the occipital bone was united with the atlas; in two others the axis was united with the third cervical vertebra, and in one, two adjoining dorsal vertebræ were soldered together. The lines of union in these cases were even and smooth and there were no exostotic growths adjacent to suggest the existence of inflammation. In one other case of union of the axis and third cervical vertebra, more doubtfully of congenital origin, there was partial destruction of the posterior arch of the axis, apparently due to suppuration.

There were four cases of union of the first and second pieces of the sternum, showing the usual incompleteness by the small cavity in the articulation. There were also some cases of fissure of the lower part of the sternum, and the shape of the ossified portion of the ensiform appendix varied as usual.

One rib was bifurcated anteriorly.

The tibiæ and fibulæ on both sides in H. 90 exhibited a marked uniform symmetrical anterior curvature. The index of the right tibia, as shown in Table LXXIV, was 53.03, a very low index, yet exceeded in this respect by several of the series. The index of the left tibia was not computed, as the bone was so split that it was feared the normal dimensions could not be obtained. These were the only leg bones that showed this curvature to any noteworthy degree. They were better entitled to the name of saber bones than any in the series. There was no certain sign of inflammation or degeneration in these bones. The skeleton unfortunately was quite incomplete, but what remained showed the following lesions: A healed fracture of the outer third of the right clavicle; small exostoses on the articular surfaces of the condyles of the lower jaw; bony growths on the sites of many tendinous insertions; a few of the vertebral bodies were very friable and a large osteophyte bound them together anteriorly. This is the only case which suggested the possibility of rickets, but the symmetry of the curvature disposes one to doubt that this disease existed.

Still it is possible that the case comes under the class described by Agnew[*] as "mild form of rickets."

Injuries.—There were some specimens showing the healing of fractures, three of the clavicle and one of the tibia; the latter had healed with marked deformity. Recent fractures could not of course be recognized because of the general fragmentary character of the bones.

In one case, where unfortunately most of the vertebræ were absent, one of the dorsal vertebræ, apparently the eleventh, showed the condition somewhat like that seen in cheesy degeneration and caries of the vertebral bodies. The body of the bone was shaped like a truncated wedge (*vide infra*). In the same case three ribs, apparently the sixth, seventh, and eighth, right side, showed posteriorly from the head to the angle a rough surface with exostotic growths, as if the ribs had formed the wall of an abscess. There was also an impacted fracture of the neck of the right femur, and the right ulna and both fibulæ showed an uneven surface that might possibly have resulted from a contusion with consequent inflammation. Altogether I regard the condition as one general injury, probably from a fall on the right side.

There was one case of fracture of rib with good union and no deformity except a slight overlapping.

There was one case of anchylosis of astragalus and *os calcis*, and another of the second metatarsal and middle cuneiform bones, both probably traumatic.

Disease.—In about one-third of the cases periosteal fringes of new bone were found along the edges of the bodies, and sometimes of the laminæ of the vertebræ. In view of the incompleteness of the individual sets, it is impossible to state with any accuracy the relative frequency with which the disease occurred in the different regions. Apparently it affected most frequently the lumbar region, next the dorsal, then the cervical, and least of all the sacral. In two cases only there

[*] AGNEW: Principles and Practice of Surgery, Philadelphia, 1878; Vol. I, p. 1032.

were actual bridges of bone, and these connected adjoining lumbar vertebræ. These exostotic growths resembled those seen in the bones of individuals who have worked hard and been exposed to cold and wet, those often found in the bones of the dissecting-room subjects. The condition may be termed "rheumatoid." It is worthy of observation that the vertebræ were much more frequently affected than the other bones of the skeleton.

The frequency of this rheumatoid condition in the people represented by these bones may seem rather surprising in view of the mild character of the climate at the present day, which is probably similar to what it was in their time; but, granting the existence of this condition, it is easy to understand its predominance along the spine. The Saladoans were a hard-working people, whose labor was of such a character as to cause much bending of the back, to make them perspire freely, and to subject them to sudden changes of temperature while perspiration was active. In short, they lived in many respects under conditions similar to those of our own laboring classes, and we need not wonder that they suffered from similar maladies of the vertebral column.

There was one case of antero-posterior curvature of the spine in an adult which merits special description. In the dorsal region a number of adjoining vertebræ had their bodies symmetrically and bilaterally diminished from behind forwards; they had the shape of a truncated wedge with its base posterior. No distinct evidence of caries could be discovered, as in Potts disease, but the friable and injured condition of the bones did not permit us to announce a positive opinion on this point. The change in shape seemed more probably due to an interstitial absorbtion than to caries. We should hesitate to say that it was a case of tubercular degeneration; there was no satisfactory evidence of the existence of such a condition in any bone in the collection. There were many fringes of new bone along the bodies of the diseased vertebræ, and there was firm coössification of adjoining bones at the left sacro-iliac synchondrosis. There may be other skeletons in this series which had similar lesions, but the loss or destruction of some of the vertebræ forbid us to speak with certainty. We have in the general collection of the Museum a skeleton from Alaska showing a condition similar to that described, and we will anticipate Part II of this work by saying that we have another such skeleton in the series of Cibola.

A disease exists in Zuñi which Mr. Cushing, freely translating the Zuñi name, calls the "warps." It consists of a gradually increasing, symmetrical, antero-posterior curvature of the spine, which, when it reaches completion, after years of progress, brings the knees in close proximity to the chest and renders walking impossible. The patient is obliged to go around on short crutches and is reduced to a helpless condition, his only useful occupation being the knitting of stockings. The disease is not accompanied by abscesses or sinuses, and the general health of the afflicted person is not seriously impaired. It is said that on the first appearance of the malady, if the patient will permit himself to be tied night and day to a straight board, he may avoid the worst consequences; but either this is not an infallible remedy or there are some who have not the fortitude to submit to it, for the writer has seen at least half a dozen sufferers in the pueblo of Zuñi, all adults and mostly males. The connection, if any exists, between this disease and the spinal curvature of the Saladoans and Cibolans, referred to above, is worthy of investigation.

In several cases the conditions suggested the possibility, but by no means demonstrated the certainty, of syphilitic disease. Thus in one there was irregular nodular hypertrophy of the shafts of both tibiæ, more especially the right, of the lower part of the right fibula, and of the shafts of both ulnæ, while the sternal ends of the first ribs showed exostotic growths. In some cases there was hypertrophy of the tibial shafts without any other evidence of disease.

The fragmentary and worn conditions of the skulls interfere with the recognition of disease and injury. There were, however, abundant evidences of alveolar abscess, more especially in the lower jaw; and in a few cases the alveolar wall was perforated. In one case the left lower incisor and part of the alveolus were absent, probably from abscess or injury, but in this situation giving a very peculiar appearance to the jaw.

In 2 or 3 cases the eminentia articularis was eroded on one side, and the corresponding condyle was also largely destroyed. It seemed to be rather the result of atrophy than inflammation.

The lesions of the jaws and teeth are father considered in the section on teeth.

§ 6. THE CEPHALIC INDEX.

The most notable feature of this collection of skulls—the feature which at once attracts the attention of the observer—is the antero-posterior shortening. Excluding those which bear un doubted evidence of post mortem distortion, the longest skull (H. 23, Table II) in 48, whose indices are obtainable has a cephalic index, computed from the measurements prescribed by Broca, of only 78.40, which is within the limits of mesaticephaly. There are but 4 skulls which are longer than sub-brachycephalic, and but 7 which are not truly brachycephalic. The shortest skull has an index of 97.97. The mean of all the 48 indices is 88.47, which is an extreme grade of brachyce- phaly—the brachystocephaly of Huxley.

§ 7. OCCIPITAL FLATTENING.

Associated with this shortening of the skull we find more or less depression or absolute flat- tening of the occiput. In the most marked cases we can not doubt that this flattening is artifi- cially, although not necessarily intentionally, produced. Between the very flat occiputs and those which, though not prominent, are quite rounded there are many degrees of variation and the areas of the flattening are of various sizes from those that comprise the entire occipital region, and show definite boundaries to those which might easily escape the attention of the student, or might, dis- associated from the rest of the group, be regarded as normal peculiarities.

The occipital flattening here referred to, must be carefully distinguished from that produced intentionally by the ancient Peruvians, by the Flatheads of our Northwest coast, and by other races. In the latter there is an anterior counter-flattening produced by the pressure applied to the forehead; in the former there is no frontal flattening.

The cause of this flattening of the occiput, whatever it may be, would seem to be the cause, under modifying circumstances, of the brachycephaly in general, whether absolute flattening exists or not. Such flattening has been observed among various American races, both extinct and extant, and is by some attributed to the use of a hard board for the back of the basket, case, or cradle in which the baby is carried.* There is no doubt in our mind that this is the prime cause of the flattening and the brachycephaly in the skulls of this series. The variations may depend on the different degrees of hardness of the skulls or on the character and size of the pad or pillow used, or on both.

In 46 cases where the occipital depression notably affects the sagittal circumference we have it variously distributed. This distribution may be broadly divided into three groups. These are illustrated in Figs. 26, 27, and 28, which consist of superimposed outlines adjusted on the bregma, the superior margin of the meatus auditorius, and German horizontal plane.

In the first group (A) the depression, whether there is actual flattening or not, is pretty evenly distributed over the entire posterior portion of the sagittal curve from the opisthion to the obelion. This may be called total posterior depression. (See Fig. 26.)

In the second group (B) the depression is mostly from the inion to the obelion; that portion of the median line below the inion seeming to be little affected, this we may designate as depres- sion above the inion. (See Fig. 27.)

In the third group (C) the flattening or depression is mostly above the lambda, the median line below that point being nearly or quite normal. (See Fig. 28.)

As the outlines of all the skulls which satisfactorily illustrate the sagittal depression have been used in composing Figs. 26, 27, and 28, it will be seen that the first group is much the most numerous, there being 14 skulls of this group to 6 of the second and 4 of the third.

---

* CARR, LUCIEN: Observations on the Crania from the stone graves in Tennessee. Eleventh Annual Report of the Peabody Museum of Archæology and Ethnology. Cambridge, Mass., 1878, pp. 361–384. SHUFELDT; A Navajo Skull. Journal of Anatomy and Physiology, Vol. XX, London, 1885–'86, pp. 426–429. SHUFELDT: A Skull of a Navajo Child. Journal of Anatomy and Physiology, Vol. XXI, London, 1886, pp. 66 et seq. SHUFELDT: Contributions to the Comparative Craniology of the North American Indians. Journal of Anatomy and Physiology, Vol. XXI, Lon- don, 1887, pp. 525 et seq. MASON, O. T.; Indian Cradles and Head Flattening. Science, Vol. IX, No. 229, New York, June 24, 1897, pp. 617 et seq.

These groups are believed to be not devoid of significance, and will be made the basis of future comparative study of American races. It has been already found, for instance, that in skulls dug from American mounds, where occipital flattening is often encountered, that total posterior flattening (Group A) is much the rarest form. In 68 mound skulls, with sagittal depression, the groups are distributed as follows: Group A, 7; Group B, 51; Group C, 10. It may be, too, that our future studies will compel us to establish another class, in which the depression is below the lambda.

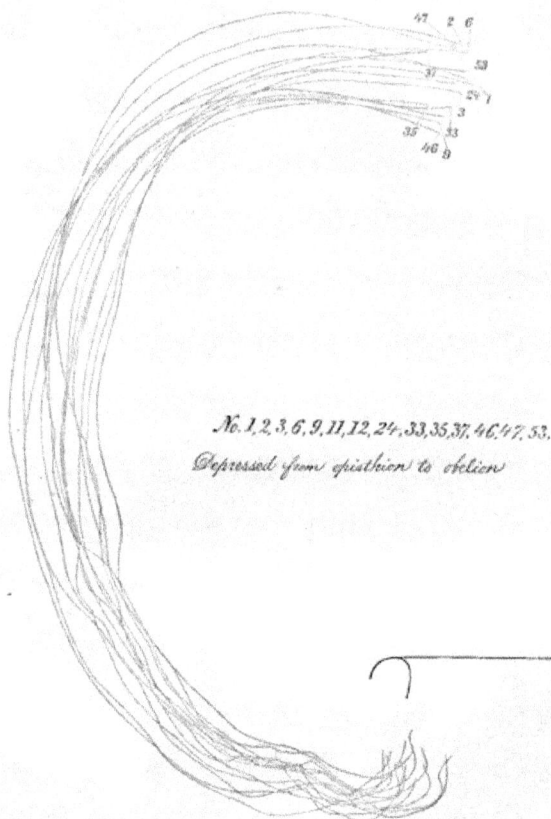

No. 1, 2, 3, 6, 9, 11, 12, 24, 33, 35, 37, 46, 47, 53.

Depressed from opisthion to obelion

Fig. 26.—Occipital depression. Group A.

Besides these sagittal variations in depression, we have different forms and **degrees** of lateral depression, i. e., the depression, instead of having its center on the median **line**, has it more or less to one side. This character naturally divides itself into two groups—right and left. In forming these groups we have depended upon a mere inspection of the skull and not upon measurements, only a skull which had an obvious lateral deformity being included in either group.

These lateral deformities are not to be profitably considered under the head of plagiocephaly as defined by Broca. According to this author a certain depression of the frontal bone on one side accompanies a depression of the parietal on the opposite side in the condition to which he applies this name. While in this collection there are some true plagiocephalic skulls, the majority having the posterior lateral flattening have not the accompanying frontal flattening. Hence they have been all first studied together with regard to the posterior flattening only. Of a skull thus

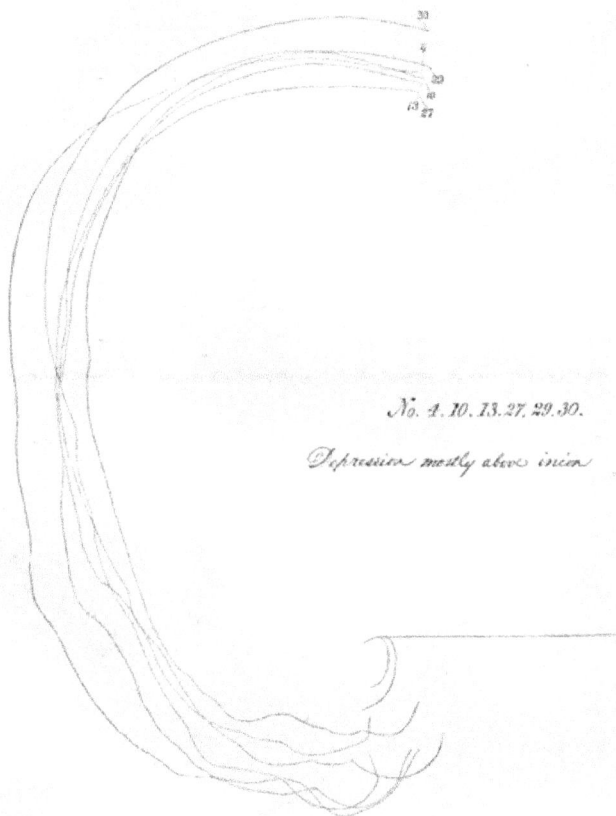

No. 4. 10. 13. 27. 29. 30.

Depression mostly above inion

Fig. 27.—Occipital depression. Group B.

flattened on the left side we say it has left posterior flattening, although it may have right plagiocephaly, and of a skull thus flattened on the right side we say it has right posterior flattening, although it may have left plagiocephaly. In short we study this deformity first without regard to plagiocephaly.

Out of 28 skulls showing the lateral posterior depression, 19 are flattened on the left side and 9 on the right. These deformities are illustrated in Figs. 29 and 30, which represent superimposed outlines adjusted on the median line and the maximum occipital point.

Thus we see that the skulls flattened on the left side are twice as many as those flattened on the right. Right-handed women carry the child usually on the left arm, and therefore suckle it mostly at the left breast,* and right-handed people predominate greatly over the left-handed in all

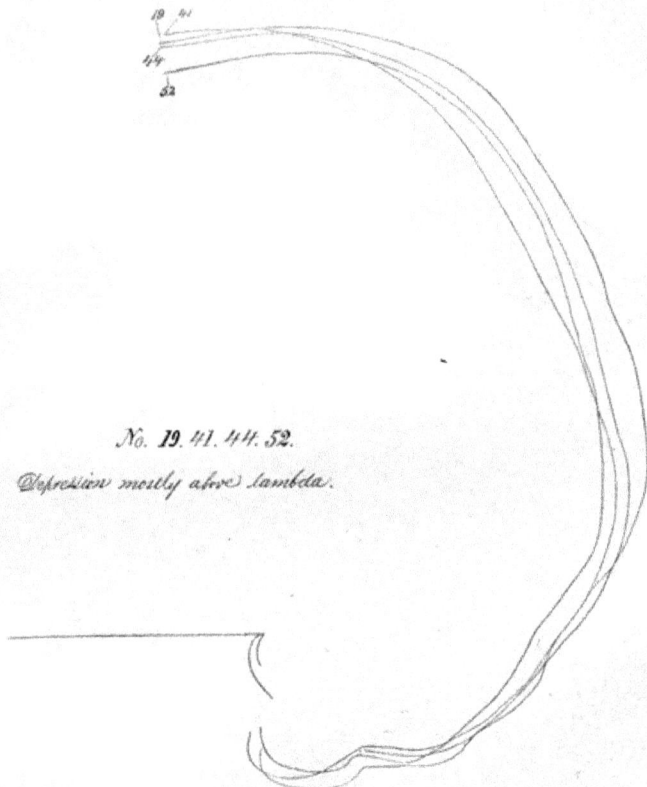

No. 19. 41. 44. 52.

Depression mostly above lambda.

Fig. 28.—Occipital depression. Group C.

races. When lying on the hard cradle-board, then, the heads of the great majority of infants should more frequently incline to the right than to the left, and should therefore, we would suppose, be more likely to become flattened on the right side. In our mound skulls the flattening is much more frequent on the right side than on the left, in the proportion of 62 right to 39 left in 101 skulls in which lateral posterior flattening is found.

* JOHN A. WYETH. "Anatomical Reasons for Dextral Preference in Man." Annals of Anatomical and Surgical Society, Brooklyn, N. Y., Vol. II, 1880, p. 129.

One peculiar effect of the occipital flattening is observed in the horizontal circumference. In certain of these skulls (H. 4, H. 10, and H. 47) a curious difficulty has been encountered concerning the horizontal circumference. It is prescribed that this circumference, which is supposd to be the maximum, be taken on a line passing above the supraciliary ridges and through the maximum occipital point; thus the posterior segment of the circumference encircles, so to speak, the posterior end of the maximum length. But in these skulls the line indicating the greatest circumference passes high up toward the obelion, and is drawn through so high a plane of the skull that the greater breadth of the skull at points below that plane more than compensates for its slightly less length; therefore the maximum circumference does not lie in the same plane as the maximum length.

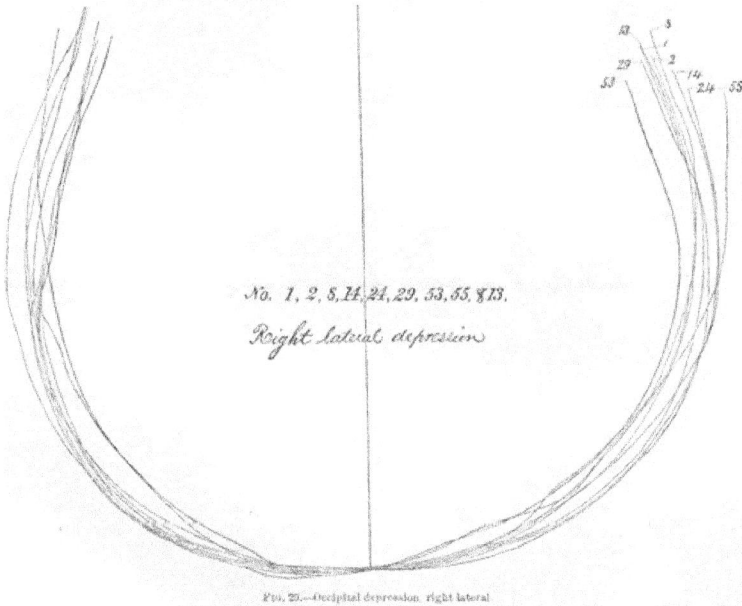

No. *1, 2, 8, 14, 24, 29, 53, 55, & 73.*

Right lateral depression.

FIG. 25.—Occipital depression, right lateral

Again; suppose that we take a skull of any ordinary shape and paint a line around it in the horizontal plane of its greatest length. If we then look downward upon the vertex of the skull we shall hardly see the line at all, because it corresponds so nearly to the outline of the skull in *norma verticalis;* but if we take one of the deformed skulls in question and paint a line correspondingly related to the maximum length and then look down upon the skull, we shall see painted upon it an ovoid figure which coincides with the outline of the skull only at its posterior extremity. This is owing to the fact that the most protuberant regions of the cranial parietes are situated much below the horizontal plane of the greatest length.

In these cases both the maximum circumference and the circumference around the maximum occipital point have been recorded, although it has been a matter of great difficulty to determine exactly the maximum circumference, and a series of measurements of the same made at long intervals of time would probably show considerable variation.

§ 8. APPARENTLY NORMAL SKULLS.

There are 16 skulls which, if never seen in connection with the rest of the collection, might readily be regarded as normal skulls. Taken by themselves, the fact that they are deformed is not obvious; studied along with the rest of the group, where there is every gradation from the most unquestionably flattened to the apparently normal, the observer has no doubt that the causes which operated in distorting the former class have had their effect too in shaping the latter, and he feels uncertain where, in any shortened skulls, he is to draw the dividing line between the normal and the abnormal. To what extent do the pillow and cradle of civilization affect the skull? In our great collection of Indian crania, those which are the longest, without obvious artificial deformity, and those which have the best developed occipital shells belong to tribes which

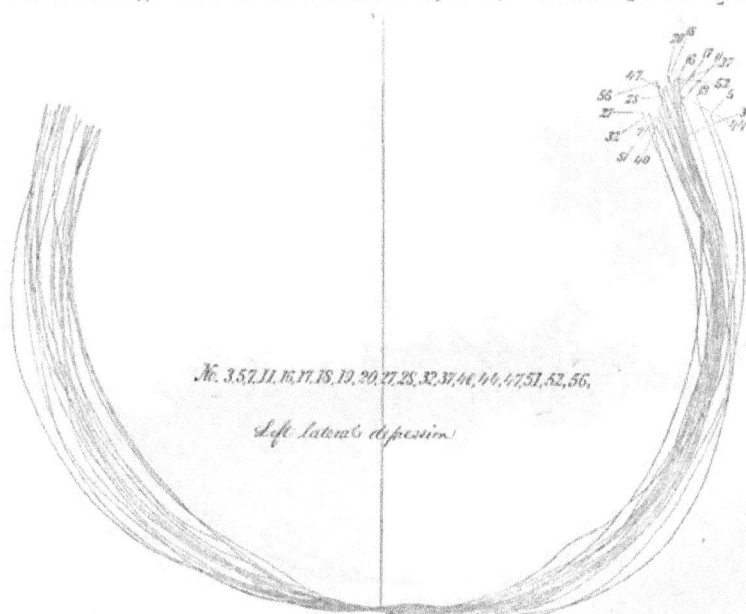

No. 3, 5, 7, 11, 16, 17, 18, 19, 20, 27, 28, 32, 37, 44, 46, 47, 51, 52, 56.

Left lateral depression.

Fig. 36.—Occipital depression, left lateral.

use no cradle-boards or baby baskets; but carry their children in soft bundles, on the back, in blankets or in frames which present a flexible surface of stretched cloth or buckskin to the occiput of the infant.

It is evident (see Tables IV and V) that these apparently normal skulls partake fully of the brachycephaly of the whole group. They represent neither the longest nor the shortest of the entire series; their extremes being 78.40 and 94.66, and their average cranial index (86.94) is but little lower than the average of all (88.47).

Many craniometricians advise that the deformed skulls like most in this collection should not have their cranial measurements taken or placed on record for comparison. Such advice has not been followed here. All that do not show decided post-mortem distortion have been measured. This **has been done because of** the uncertainty referred to above in distinguishing between the normal and the abnormal, because the occipital distortion is found in the skulls of so many of our American races, and because it is felt that its careful study by measurement may eventually prove of great value in comparing the races. In some cases, however, separate tables have been arranged for the apparently normal skulls, which are designated as follows: Numbers H. 7, H. 12, H. 15, H. 18, H. 19, H. 21, H. 23, H. 25, H. 26, H. 34, H. 36, H. 39, H. 40, H. 44, H. 54, H. 57. (See Tables IV and V).

§ 9. POSITION OF MAXIMUM OCCIPITAL POINT.

A feature, probably the effect of occipital distortion, which is usual in these skulls is the elevated position of the maximum occipital point. In 50 specimens in which the position of the lambda may be determined, we find that the maximum occipital point lies above it in 10, and at it or less than 5^{mm} below it in 10 more. In other words, the maximum occipital point lies without the occipital bone in 20 per cent of the specimens and is barely included in the latter in another 20 per cent. In 3 of the former 10 skulls the point is seen in the region of the obelion. In the remaining 30 skulls, while it is found on the occipital bone, it is usually found high on it. In only 5 cases (10 per cent of all) does the point appear in the region of that usually ill-defined locality, in these skulls, the inion.

§ 10. THE LENGTH-BREADTH INDEX.

The equality in this collection of the cephalic index of Broca to the length-breadth index of the Frankfort agreement is remarkable and is due no doubt to the occipital flattening. The maximum occipital point being unusually elevated by reason of the flattening (§ 9), it often coincides, or nearly coincides, with the posterior extremity of the German horizontal length, thus approximating the only factors of these two indices that differ. In 13[*] out of 47 cases these two indices are exactly equal to one another; in one-half[†] of the 34 remaining cases the indices differ less than one unit. According to this index the longest skull is again H. 23, and it is one of those skulls in which both indices are alike. H. 46 is again the shortest skull, but its horizontal length being shorter than its greatest length, we have the higher length-breadth index of 99.31. According to the "agreement" concerning this index, 3 skulls only are mesocephalic, 8 are brachycephalic, and 36 are hyperbrachycephalic. The average, closely approximating that of the analogous vertico-transverse index, is 88.75, an extreme grade of brachycephaly.

In 10 instances[‡] we have the confusing record of a vertico-transverse index higher than a length-breadth index. This involves the paradox of a length greater than the maximum length. A reference to measurements 6 and 7, in Table I, will show, furthermore, that such is our actual entry in the cases where footnotes are referred to. This apparent inconsistency arises from the following conditions: First, the occiputs of these skulls are so distorted that one side of them projects posteriorly beyond not only the other side but beyond any point in the posterior part of the sagittal plane, so that the profile of the skull does not correspond in outline to a section in the sagittal plane. Hence, second, the longest dimension parallel to the horizontal plane is not in the sagittal plane. We do not, however, measure directly from the glabella to the most prominent side of the occiput, which would give us an oblique measurement, but by means of the vertical plates of Spengel's craniometer we measure that which is a line parallel to the sagittal plane but lying to one side of it. Imagining this line to be projected upon the sagittal plane, we reckon our indices according to the accepted formula. We might have so modified the results or the modes of measurement as to remove this discrepancy from the record, but we considered it more candid as well as more scientific to give the results as originally determined.

§ 11. THE VERTICAL INDICES.

The occipital depression referred to not only directly shortens the antero-posterior diameter, but increases the height and width of the skulls actually as well as comparatively. As a consequence, not only is the cephalic index very large throughout the group, but the vertical indices are correspondingly exaggerated.

---

[*] Nos. H. 3, H. 5, H. 6, H. 7, H. 11, H. 12, H. 14, H. 18, H. 23, H. 44, H. 52, H. 55.

[†] Nos. H. 9, H. 13, H. 15, H. 16, H. 21, H. 25, H. 26, H. 27, H. 34, H. 35, H. 36, H. 37, H. 41, H. 42, H. 50, H. 53, H. 56.

[‡] Nos. H. 9, H. 19, H. 21, H. 25, H. 28, H. 40, H. 42, H. 53, H. 56, H. 57.

The vertico-longitudinal index was obtained in 40 skulls (Table VIII). Its extremes in adult skulls are 78.79 and 97.29. We have a child's skull, however, which has an index of but 77.70, and it is well to observe that the maximum index (skull H. 32) is far removed from the rest of the group, the next greatest being 92.56. The average of 39 adult skulls is 83.24.

Our lowest index is within the limit of high skulls as given by any known authority. Sir William Turner applies the term acrocephalic to all crania with an index of 77 or above.[*]

In the list of the vertico-transverse index (which may be computed in 39 skulls) the lowest is 84.82, the highest is 105.88. This belongs to the same skull, which has the highest length-height index, namely, H. 32. In respect to the index now under consideration, H. 32 is not so far removed from the rest of the group as it is in the length-height index, as will be seen in Table XIV, where the highest five indices are: 105.88, 104.47, 103.62, 102.27, and 101.39.

In 36 skulls both the vertico-longitudinal and the vertico-transverse indices were ascertained, and from those we were able to determine the *mixed index of height* of Topinard.[†] Of these 36 skulls the average vertico-longitudinal index is 85.40, the average vertico-transverse index 96.49, and the mixed index 90.94. (See Table XVII.)

A casual glance at the above figures might lead to the conclusion that the pressure on the occipital tended more to increase the width than the height of the skull, but such is probably not the case. The transverse measurement is taken wherever the maximum width falls; the height measurement is taken from basion to bregma, and the latter is in no case the highest point on the sagittal suture in this group—it rarely approximates the highest point. If a series of vertical measurements were taken from either the German horizontal plane or the alveolo-condylean plane extended, the most distant point of the sagittal suture would usually be found posterior both to the bregma and the vertex of Broca, and often nearer to the obelion than to either. Thus it probably is that the vertico-transverse index is the greater of the vertical indices.

§ 12. THE PLANE OF THE *FORAMEN MAGNUM* OR OPISTHIO-BASILAR PLANE.

In 29 skulls, where the landmarks were intact, we have determined the degree of inclination of this plane according to the three methods usually employed, i. e., we have taken the angle of Daubenton, the occipital angle of Broca, and the basilar angle of Broca. Tables XVIII–XXIII give the results of our measurements, recording in no case less than half a degree.

We are told[‡] of the angle of Daubenton that its lowest recorded expression is − 16 in an Auvergnean, and its highest + 19 in a Hottentot. In the Hemenway collection we have no minus quantities for this angle; our lowest is + 4° 30′, while our highest far exceeds this exemplary Hottentot, being + 23°. The highest average we have seen mentioned is + 9.34 in Nubians, but the average of the Saladoans is 13.30.

The occipital and basilar angles of Broca are, of course, correspondingly exaggerated in our series, the mean and extreme of the former being respectively 24° 15′ and 35°, and of the latter 32° 15′ and 46° 30′ (Tables XX and XXII). The mean of the Nubian basilar angle is 26° 32′.

The opisthio-basilar line is very approximately a continuation of the alveolo-basilar line in skulls H. 10 and H. 23, whose angle is, of Daubenton, 18° 30′. A straightedge applied to the median line at the base may be made to almost touch at the same time the alveolar point, the basion, and the opisthion. We may say, then, that the plane of the foramen magnum in these two cases looks directly downward. In skulls H. 18, H. 24, and H. 25, whose angles exceed 18° 30′ the plane looks downward and backward. In the rest of the series it looks downward and more or less forward.

If the inclination of the plane of the foramen magnum were accepted as a measure of evolution, the Saladoans would stand at the bottom of the human scale. We are inclined either to regard their peculiarity in this respect as additional evidence in support of Topinard's opinion that the

* TURNER: The zoölogy of the voyage of H. M. S. *Challenger*, Part XXIX, Report on the Human Skeletons—The Crania. London, 1884, p. 5.
† *Op. cit.*, p. 683.
‡ BROCA: Sur l'angle orbito-occipital, Revue d'Anthropologie, t. 6, Paris, 1877, p. 334.

character which the angle of Daubenton expresses is not of "a serial anthropological character,"[*] or to think that the pressure on the occiput, before referred to (§ 7), has influenced the position of the *foramen magnum*. It is not, however, in the most flattened skulls that we find the highest angles.

§ 13. CAPACITY OF CRANIAL CAVITY.

We have already stated that it was found necessary to repair the greater part of the skulls of this series with plaster of Paris. This often so encroached on the cranial cavity as to make it impossible to determine the cubic capacity of the latter. In 8 skulls only were we able to find the cubic contents of the brain case, and these were so friable that neither water nor shot could be used in them. The measurements were made by means of mustard seed; not according to any method previously laid down for the use of this seed, but by a system of our own, approximating closely to Broca's method for the use of shot.

The plan is as follows: Use the funnels, rammer, and tin vessels as for shot cubature. Use the 2,000-centimeter graduated glass with its leveling rammer. First, lay the skull on its vertex. Pour in rapidly a liter of seed through the wide-necked funnel. Pour in, in same way, so much more seed that when the skull is set upon its face and frontal bone the seed will form nearly a level across the skull from foramen to near middle of sagittal suture. Second, insert large end of rammer into foramen, gently press seed toward frontal region, with side of rammer in such a way as to level the surface of whatever quantity of seed is in the skull. Third, fill small-necked funnel with seed; hold it in left hand with finger over its mouth. The skull, as has been said, is standing on its frontal region. Grasp occiput with right hand and slowly incline the skull into the vertex-downward position as before, while running in seed through small funnel. During this operation the seed will overflow the foramen three or four times; when it does so, thrust seed into skull with forefinger of right hand; but as soon as pressure is felt, stop pressing. When the finger can no longer be introduced into the foramen without feeling decided resistance, and the skull has been completely lowered into vertex-down position, let the seed form a heap over foramen, press this heap vigorously into foramen with right thumb, and add seed to level foramen. Fourth, cover foramen with cotton wad and shake stray seeds from surface of skull. Fifth, empty contents of skull into double liter tin. Pour all the seed as rapidly as may be from double liter into 2,000 c. c. eprouvette, using no funnel. Bring leveling rammer of eprouvette down firmly, but not violently, on seed to level it. Sixth, read the eprouvette.

That the above method gives good results which fairly admit of comparison with results obtained by shot, there is little doubt, for the following reasons: We made according to this plan five measurements on one of Professor Ranke's bronze skulls, which was presented by the inventor to our Museum. The capacity of this bronze cast, as ascertained by ourselves, with water, atmosphere, and all accessories at a temperature of 60° F., was 1,240 c. c. (The bronze is marked 1,250 c. c., but this seems to be an error.) Our measurements of the capacity of this object with mustard seed ranged from 1,239 + to 1,250 —, with an average of 1,242. We made measurements on this plan of some natural skulls in our collection which had been repeatedly measured with water and with shot according to Broca's system, and, applying three or more measurements to each skull, we arrived at results more uniform than those obtained with the artificial skull of Ranke. The mustard seed gave as a rule higher figures than those obtained by shot or water, but the average excess was less than 1 per cent.

Of the eight skulls measured four were male, two female, and two of doubtful sex. With such a small series we have not considered it proper to study the capacity of the sexes separately. The highest capacity belongs to a female skull, the lowest capacity to a skull of unknown sex. All the specimens pertain to subjects of mature age, and none are senile.

The highest two capacities are, in cubic centimeters, 1,530 and 1,510, which according to Broca's nomenclature† belong to skulls of the medium or ordinary class. Four capacities, viz, 1,390, 1,330, 1,310 and 1,170, belong to his class of small (*petite*) skulls. The remaining two capacities, 1,150 and 1,120, belong to the microcephali or lowest class. The average capacity is only 1,313. It might be urged that since our series of capacities represent such a small proportion

_____
* Op. cit., p. 814.   † TOPINARD, op. cit., p. 610.

of the whole series, this average may be far from the true average capacity of all; but we have the following reason for thinking otherwise: That capacity (1,330) which comes nearest to the above average, belongs to skull H. 7, and this it is, that next after the "type" (H. 40) is the most typical skull of the whole series as shown by its various indices.

The table of Broca, with which we have to compare this average, gives separate figures for the males and the females of each race. We have calculated the combined averages and made our comparisons with these. Broca gives 29 series including the most diverse races, but no American Indians. We find but three of his series having a lower cranial capacity than the Saladoans; these are the "Hottentots and Bushmen," the "Australians," and "Parias of Allipoor (Calcutta)." Such inferior races as the Negroes of Africa, the Papuans, the New Caledonians, and the Tasmanians seem to rank in this character above the Saladoans.

Our Table XXIV presents some small series of average cranial capacities of (lower) races represented in the Army Medical Museum. They are taken from the series of 101 (see Table LXXXI)—2 Navajos and 10 Peruvians being added. All the races, not American Indians, mentioned in this table, viz, Sandwich Islanders, Mongolians, New Zealanders, American Negroes, and Eskimos, it will be seen, have larger brain cases than our Saladoans. The position of the latter with regard to other autochthones of both North and South America is shown in the following extract from the table:

(1) Siouan tribes.................................... 1463
(2) Pah Utes 1367
(3) Apaches 1331
(4) Ancient Californians.......... 1323
(5) Navajos. 1315
(6) Saladoans 1313
(7) Peruvians 1295

It is not in accordance with current theories that a people as advanced in arts and social organization as that of the Salado Valley evidently was should have a cranial capacity superior only to such low races as the Hottentots and Australians. It must be borne in mind, too, that the uncremated remains of the Saladoans probably represent a superior class of this community. Still, small as is their cranial capacity, it is greater than that of the Peruvians, who were a race more advanced than the Saladoans. We have little to suggest in explanation of these facts. Perhaps the subject of cranial capacity in relation to culture may have to be reconsidered. The Saladoans were a people of low stature and rather slight physique, and the relation which the skull bears to the rest of the skeleton may be a factor in the problem. We have as yet no evidence to show that distortion reduces the capacity of the cranium.

§ 14. THE TYPE SKULL

The following method is the one we have adopted for selecting a type skull from the series: First. Let all the sets of indices be arranged in ordination. Second. Subtract the lowest index in one ordination from the highest. Third. Divide the difference by 2, and add the quotient thus obtained to the lowest index. This gives the theoretical mean of variation. Example: Suppose we have a series of skulls with cephalic indices ranging from 80.00 to 90.00. The first step, subtraction, gives us 10.00; the second step, division, gives us 5.00, and the third step, addition, gives us 85.00, which is the theoretical mean of variation. The skull, if any, having this index is the type of the series as far as concerns the cephalic index. In practice, however, where we calculate indices to the second decimal place, it is not usual to find any skull with the index exactly expressing the theoretical mean. The skull most nearly expressing it is taken as the type.

It follows that if we take many different series of indices upon the same skulls we have to determine what skull stands in the plurality of instances nearest the theoretical mean. Suppose we calculate ten different series of indices upon 9 skulls (an odd number is easier for the purpose of explanation). If 1 particular skull expresses the theoretical mean of variation in each and every series. it is, of course, the type skull of the lot in every respect, so far as the investigation has gone.

But if, as must always be the case, no skull expresses the theoretical mean of every series of indices, then we take the skull which averages nearest the theoretical mean. Therefore, of our supposed lot of 9 skulls we select, let us say, from the first series of indices, 3—the skull having an index most nearly expressing the mean of variation, the skull having an index next greater than this, and the skull having an index next smaller. Now, supposing that we have ten series of indices, let us say that the skull which expresses the theoretical mean of the first series comes nowhere near it in any other series, while the skull next below the theoretical mean in the series of indices under consideration is the theoretical mean of two other series of indices, and stands either just above or just below the mean in every series. The latter, then, is very likely the type sought.

To state it more methodically: We have measured a lot of skulls, have reckoned their indices, and have arranged the several different kinds of indices in as many different ordinations. In each ordination we select the index most nearly expressing the mean of variation and call it No. 1; the index next above and index next below this we call No. 2. The index next above the greater No. 2, and the index next below the lesser No. 2 we call No. 3, and so on. Now let us add together, for each skull separately, the Nos. 1, 2, 3, etc., expressing the position of the several indices with regard to the theoretical mean of radiation of each series of indices. Divide the sum thus obtained by the number of series of indices. The skull whose indices thus treated give the lowest quotient is the type.

In the present case, however, it must be remembered that the Salado skulls are much broken, so that only a few can yield a complete series of measurements. The type skull, therefore, in part owes its selection to its good preservation, it being represented in every series of indices. It can not be said to be the type of 57 skulls, perhaps, but in a general way, all things considered, it is the best representative of the characteristic dimensional relations of crania of the people in question. Its most aberrant feature consists in the unusual height of the orbits, shown by the orbital index 96.05, while the theoretical mean of the orbital indices is 90.90.

The type skull thus selected is H. 40; its five views are shown in plates 55 to 59, inclusive. Of skulls in good condition H. 7 and H. 15 approach nearest to the type.

§ 15. PROCESSES AT BASE OF SKULL.

There is evidence, in the archæologic find of Los Muertos and Las Acequias in the shapes of the pottery, etc., that this people, like the modern Pueblos, were accustomed to carry heavy burdens on the head. Such being the case, we might reasonably expect to find the various processes for muscular and ligamentary attachments at the base of the skull strong and prominent; but, on the contrary, we find them unusually subdued and weak. It may be that our expectations are unfounded; that the load on the head, once well balanced, required little muscular exertion to sustain it.

The Inion.—In 46 adult skulls, with this process well preserved, compared with the five forms of Broca,[*] we find that 27 agree with his zero or lowest form, that 19 resemble his No. 1, and that none are to be considered of a higher grade than this. It has been conjectured that the general pressure which has flattened the occiput in these skulls may have hindered the full development of the inion; but the fact that all the processes of the base are weak, and that the inion is ill developed in skulls where the pressure did not fall upon it, seems to indicate that pressure can at most account for only a part of the subdued features of the inion in this series.

§ 16. THE PTERION.

Of the pteria 32 are sufficiently preserved to be studied with profit. They occur in 24 skulls, 13 on the right side, 19 on the left. Four exhibit the character plainly, but can not be measured. The remaining 28 (see Table XXV) are easily measured. Only 8 skulls have the pteria intact on both sides.

*BROCA: Instructions craniologiques et craniométriques. Pl. VI.

All are of that form called by Broca pterion in H. Wormian bones complicate their characters.

Of the 11 measurable right pteria the longest is 20^{mm} (the maximum of the group) and the shortest is 5^{mm}. Of the 17 measurable left pteria the longest is 18^{mm} and the shortest 3^{mm} (the minimum of the group). The average length of the right pteria is 12.90^{mm}; the average of the left pteria is 11.35; the average of all, 11.96.

There are but two pteria of less than 8^{mm} in length, a percentage of 6.5, which is smaller than any on Anontchine's[*] table except that of the Peruvians, which is 3.4. There is but one pterion which does not exceed 3^{mm}, but with our small total of 28 this gives us a percentage of 3.5.

On the whole the character of the pterion is of a very high type.

§ 17. UNIQUE SAGITTAL SYNOSTOSIS.

The presence among the Saladoans of 4 skulls showing unique sagittal synostosis, one of them adolescent, has naturally led us to inquire if an early sagittal synostosis can be a physiological characteristic of this people, or if, at whatever age synostosis begins, it affects first the sagittal suture. With this point in mind we have investigated several other series of American skulls with the following results (the description applies solely to the outer table except in cases where the inner is expressly mentioned):

Saladoans.—The Saladoans present four cases of unique sagittal synostosis as follows:

No. H. 15, a fairly well-preserved skull, female; basilar suture closed; third molars cut except left lower (?); right lower second premolar and first molar shed and alveoli absorbed; right upper third molar decayed away; the two third molars still visible; lower right and upper left one only slightly worn, especially the latter; premolars and first molars worn just into the enamel. Sagittal suture completely obliterated; no other synostosis. A line of porosity across the pre-occipital may possibly indicate previous existence of an *os epactale.*

No. H. 17, a well-preserved skull of a youth; basilar suture open; all milk teeth shed; no third molars cut; no permanent teeth lost ante-mortem; first molars a little worn; entire obliteration of sagittal suture; no other synostosis of brain capsule.

No. H. 45, a fairly well-preserved female skull; basilar suture closed; full set of permanent teeth cut and none of them lost ante-mortem; wear of enamel very slight; complete sagittal obliteration; no other synostosis.

No. H. 49, a much warped and laterally flattened skull; post-mortem distortion; basi-occipital broken away; full set of teeth, except lower third molars, cut and none shed ante-mortem; lower third molars point forward and are impacted against second molars, probably never would have been erupted through gum; first molars worn, but not into dentine; obliteration of **sagittal** suture; probably no other synostosis; sutures of cranial vault all very simple.

Peruvians.—Among the Peruvians the following cases are to be noted in connection with sagittal synostosis:

No. 2315, a well-preserved skull without mandible; basilar suture closed; permanent teeth all erupted and none lost ante-mortem; all teeth lost post-mortem except left upper first premolar and molar; these teeth worn into the dentine; posterior two-thirds of sagittal suture obliterated; anterior third ossified in spots; no other synostosis; there is a slight ridge about the anterior part of the sagittal suture; the left temporal sends a process to join the frontal bone.

No. 2506, a well-preserved skull without mandible; basilar suture closed; teeth all cut, but third molars lost ante-mortem; all teeth which are present are worn down to the dentine; complete sagittal obliteration; a very little commencing synostosis of the lambdoid and left occipito-mastoid sutures; no other synostosis; sagittal ridge; a process joins right temporal and frontal.

No. 2945, a well preserved skull with mandible; basilar suture closed; all permanent teeth cut; both upper third molars and left lower third molar shed ante-mortem; teeth worn down to the dentine; complete obliteration of sagittal and complete obliteration of right squamous suture; no other synostosis; sagittal ridge.

---

[*]Anontchine; Sur quelques anomalies du crâne humain et de leur fréquence dans les races. Review by C. de Mérejkowsky in Revue d'Anthropologie, 2d series, vol. 5 (1882), p. 359, *et. seq.*

It is seen from these notes that the Peruvians offer no case of sagittal synostosis comparable to that of the Saladoans. This is the conclusion arrived at by considering that all three of the above skulls are at least mature and show a sagittal ridge.

Yucatees.—The Yucatees offer the following specimens of unique sagittal synostosis:

No. 626, a well-preserved skull without mandible; basilar suture closed; all permanent teeth cut; second upper right premolar shed ante-mortem; teeth not worn; sagittal suture obliterated; no other synostosis.

No. 628, a well-preserved skull without mandible; basilar suture closed; all permanent teeth cut; left upper third molar lost, probably ante mortem; enamel of teeth not worn; sagittal suture obliterated: a little commencing synostosis just above the lambda; no other synostosis.

Californians.—No. 1415, small, rather heavy and well-preserved; basilar suture closed; third upper molars cut (lower jaw not found); right upper third molar lost; the teeth show wear sufficient to slightly expose the dentine except in the case of the left upper third molar, of which the enamel alone is worn; the sagittal is coössified throughout its entire length on the inner table, and all but its anterior fifth on the outer table; no other synostosis.

No. 1430, medium size; facial bones separated from cranium and only right side of mandible preserved; third molars cut; but all but right lower have been lost; basilar suture closed; the enamel only of the teeth is worn; sagittal entirely obliterated; no other synostosis.

No. 1507, small skull; right temporal and cerebellar regions broken away; mandible broken across the symphysis; basilar suture closed; all third molars cut, but right upper one has been lost; the second and third molars have their enamel only worn; some other teeth have their dentine slightly worn. The sagittal is coössified entirely on the inner table and all but its most anterior portion on the outer table; no other synostosis; lambdoid quite complicated.

No. 1748, consists of the cranial vault, only, from a good-sized, rather scaphoid specimen; sagittal completely obliterated on each table; coronal and lambdoid fully open; no way of judging age.

A skull which is less satisfactory to discuss, as all its teeth have dropped out post-mortem, is No. 802, a well-preserved recent skull; basilar suture closed; third molars cut; all teeth dropped, but there is no alveolar absorption. There has been a large *os epactale*; it is now firmly coössified to the parietals, and they in turn to each other, the sutures being thoroughly obliterated; other sutures, including that between the epactal and occipital, open. Rather a scaphoid skull.

In many of the Californian skulls there is a prominence, sometimes prolonged into a ridge, just behind the bregma; none of the above synostotic skulls show this peculiarity except No. 802, which has a slight ridge. In some skulls, however, where synostosis is more general and probably a purely senile change, it is evident enough.

Mound builders.—No. 556, a mutilated skull of a Floridian without mandible; state of basilar suture indeterminable; upper set of permanent teeth all cut; right upper third molar shed ante-mortem, teeth all deeply worn; complete sagittal obliteration; no other synostosis.

No. 1110, a mutilated skull of a Floridian without mandible; basilar suture closed; teeth mostly shed ante-mortem and rear alveoli much absorbed; sagittal obliterated except at its anterior half centimeter; very slight commencing synostosis of lambdoid; no other synostosis of brain capsule. This is probably the skull of quite an old person.

No. 730, a fairly well-preserved skull from Kentucky with mandible; basilar suture open; all permanent teeth cut except third molars; no teeth shed ante-mortem; all teeth lost post-mortem except right upper first molar and both lower first molars; these teeth are not worn; sagittal suture open anteriorly for its first 18 millimeters; behind this it is obliterated to within 16 millimeters of the lambda, and the space of these last 16 millimeters is partially co-ossified; no other synostosis. As far as age and a globose appearance are concerned, this skull is essentially similar to the Saladoan, H. 17.

No. 1012, a large, well preserved skull, with mandible, from Illinois; basilar suture closed; all permanent teeth cut; right upper molars, left upper first and third molars, and both lower first molars shed ante-mortem and alveoli absorbed; teeth somewhat worn; complete sagittal obliteration; synostosis of coronal right and left between stephanion and pterion; fronto-sphenoid sutures coössified right and left; very slight commencing lambdoid coössification. This skull is interesting, not because the sagittal synostosis is unique, but because it is so complete and apparently so much in advance of the other synostoses, while it retains a globose shape, being notably rounded in its outline.

No. 1602, a well-preserved skull, with mandible, from Dakota; basilar suture closed; permanent teeth all cut and none of them shed ante-mortem; teeth show some wear; sagittal suture obliterated except its anterior two centimeters; no other synostosis.

Alaskans.—Sagittal synostosis unaccompanied by other synostoses is not conspicuous. Only the following two, the first of which is very remarkable, are worthy of note:

No. 2454, perfectly preserved, from a child between 7 and 10 years old; basilar suture open; sutures between basi-occipital and exoccipitals partly open; premaxillary suture visible on palatine vault; first permanent molars cut and second appearing; upper median, lower median, and lateral incisors cut but lost; no permanent canines or premolars; posterior two-thirds of sagittal suture entirely closed on outer and inner tables; for one-half of the remaining third there is partial synostosis on both tables, while the anterior sixth is open; no other synostosis.

No. 2486, a perfectly preserved skull of an adult; basilar suture not quite closed; lower third molars not cut; posterior two-thirds of sagittal firmly coössified; half the remainder partially so; anterior extremity open; coössification also, but less complete, of right coronal between sagittal and temporal line, right lambdoid and left parieto-mastoid.

Eskimos.—Among the Greenland Eskimos there is no case of sagittal synostosis, alone, comparable with that of the Arizonians. There is found, however, the following extraordinary specimen:

No. 1226, a skull without mandible, of light weight and well preserved; basilar suture open; teeth lost post mortem, except the right upper molars, three in number, and left upper first molar; both third molars **cut**. The one remaining in the skull is not worn, nor is the second molar much worn; dentine of first molar worn a little; complete sagittal obliteration; also complete lambdoid obliteration except about 5ᵐᵐ of the left lateral end; no other synostosis. Thus the whole posterior end of the skull from the coronal suture above to the basilar suture below is a single bone.

We conclude then: First, that unique sagittal synostosis may take place at a very early age; second, that it does not necessarily produce a scaphoid skull; third, that it may or may not be accompanied by a sagittal ridge; fourth, that at present it can not be said to be peculiarly characteristic of any American race.

Percentages representing the number of cases of unique sagittal synostosis in relation to the total number of skulls in each given series might be reckoned; but they would probably not accurately represent the tendency to unique sagittal synostosis for the following **reasons:**

First. It is most likely true that under a certain age no skull is liable to sagittal synostosis except for pathological reasons. Diseased skulls should of course be excluded from consideration and not be allowed to affect the percentage. But, inasmuch as we have learned that sagittal synostosis may take place before the skull is matured in any other respect, we must confess ourselves at a loss to determine exactly what that age is. So then we must either draw an arbitrary line between two supposed classes of skulls, the one of which is liable and the other not liable to sagittal synostosis, or we must consider every skull as liable to it. In neither case can our percentage exactly represent the facts.

Second. Cases of sagittal obliteration may occur which are striking in their completeness, but which are accompanied by very slight disseminated synostosis of other sutures. In determining whether such cases are to be allowed to affect the percentage or not, personal judgment—always a little arbitrary—must be used.

Third. Cases may occur where the sagittal is coössified but not entirely obliterated, while all other sutures are completely open. This occurrence in a young skull merits mention; but here again fallible judgment is called into play to pass upon the age of the skull and the minimum amount of synostosis entitled to mention.

In an article entitled " Nachtrag zur Anatomie der Schädelnähte," by Dr. E. Zuckerkandl,[*] in a series of 134 skulls, mostly Negroes, Negritos, and Malays, we find the following notable cases:

No. 98: Indian (American?), sagittal suture obliterated in places. No synostosis of coronal, lambdoid, or mastoid sutures.

No. 104: Peruvian, sagittal totally obliterated; other sutures open.

No. 124: Albino, sagittal obliterated in places; other sutures open.

No. 128: Javanese woman, sagittal totally synostosed; other sutures open.

In none of these cases does the author note a senile appearance of a skull, as seems to have been done when required throughout the article.

Unfortunately we do not possess a copy of Dr. Davis's work[†] on synostosis of cranial sutures where this subject is discussed. He refers to it, however, in his "Thesaurus Craniorum," which we quote:

I have * * * * * pointed out that *scaphocephalism is far from being the usual result of the early ossification of the sagittal suture*. This position is maintained by an analysis of the twenty-seven skulls in this collection which present no appearance of sagittal suture, but only four of which are true *scaphocephali*.[‡]

Continuing, he refers particularly to four of these cases of synostosis, which we quote accordingly, omitting the measurements.

No. 100: African negro, male, æt. c. 30; presents a complete obliteration of the sagittal suture but no *scaphocephalism* or other deformity. The alisphenoids and parietals only just touch.

No. 378. Pokomame; imperfect calv. * * * an instance of premature ossification of the sagittal suture which is totally obliterated. The other sutures are all open. * * * In the synostosis of the parietals in the case of a calvarium artificially deformed in an extreme degree and in a direction running parallel to the sagittal suture; it is, I believe, unique. There is not the slightest approach to scaphocephalism.

No. 915: Australian, female, æt. c. 17. This small cranium is synostotic from premature obliteration of the sagittal suture, which has not materially changed its form. It can not be denominated scaphocephalic at all. It exhibits marks of old injuries on the frontal, parietal, and occipital bones, and has no spheno-parietal sutures.

No. 789: Fatuhivan, male, æt. c. 17. This calvarium of a young subject is very large, thin, and in appearance swollen out as if it had been hydrocephalic. It is also synostotic, the sagittal suture being totally obliterated; yet the calvarium is **not scaphocephalic**, nor indeed deformed in any way.[§]

§ 18. THE INCA BONE AND ALLIED FORMATIONS.[‖]

Perhaps the most interesting feature discovered in this series is the great prevalence of the Inca bone and its kindred anomalies. This was first observed by Dr. Wortman while he was engaged in collecting and preserving the bones as they were exhumed at Camp Hemenway in the Salado Valley in 1887. He had, however, no opportunity in the field for making a careful study and determining the comparative frequency of the anomalies; besides, the bones when unearthed were in such a friable condition that they could not properly be examined until they were strengthened and repaired. Since they have been repaired at the Army Medical Museum we have found, among complete skulls and fragments, a series of 88 occipital bones in a sufficient state of preservation to be examined for these formations.

We need not enter into an elaborate description of these anomalies, nor discuss at length their morphological characters. Such elaboration is not within the general plan of our work. The accompanying illustrations will, we hope, serve to make clear to the reader, when the text may be too concise, the full meaning of the terms we employ. Those who desire to consult the original authors whom we have followed are referred to the works of Virchow,[¶] Anoutchine,[**] and Topinard.[††]

---

* Zuckerkandl, E. in Mittheilungen der Anthropologischen Gesellschaft in Wien, Band IV, 1874, p. 144 et seq.

† Davis: On synostotic crania among Aboriginal Races of Man. Haarlem, 1865.

‡ Davis: Thesaurus Craniorum, London, 1867, p. 57.

§ Davis, op. cit., pp. 105, 235, 261, and 321.

‖ Much of the material in this section has appeared previously in an article, by the author, entitled " The Inca bone and kindred formations among the ancient Arizonians." American Anthropologist, Washington, D. C., Vol. II, p. 337 (October, 1889).

¶ Virchow: Ueber einige Merkmale niederer Menschenrassen am Schädel. Berlin, 1875. Zeitschrift für Ethnologie, v. 20, 1888, p. 470.

** Revue d'anthropologie, 1883, p. 149 (Review).

†† Op. cit., p. 769, p. 791, foot note.

In the first place we will consider the true epactal bone, or *os Incæ*. It exists in all races; it becomes a characteristic of the Peruvian or Inca race only by reason of its great frequency among them. How often it is found in them more than among other peoples hitherto studied will be seen in the accompanying table (Table A).

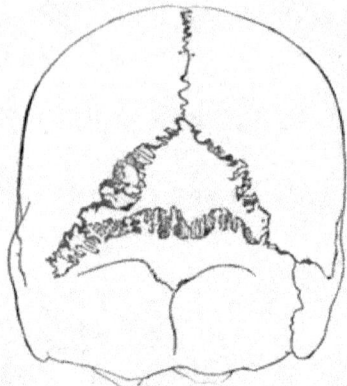

FIG. 31.—Inca bone (No. II. 13).   FIG. 32.—Inca bone (No. II. 29).

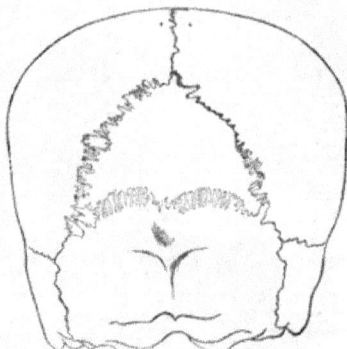

Figs. 31 and 32 represent typical forms of this bone in two varieties described by Virchow.* In one the persistent transverse occipital suture runs directly from one asterion to the other, and seems but a continuation of the parieto-mastoid suture. In the other the ends of the transverse

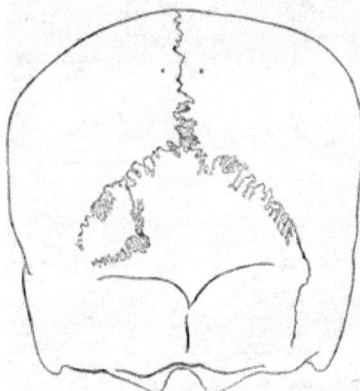

FIG. 33.—Incomplete Inca bone (No. II. 14).   FIG. 34.—Quadrate bone (No. II. 48).

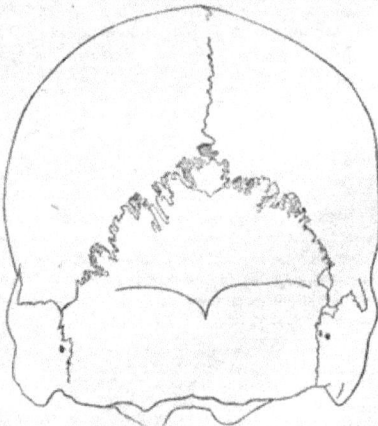

suture join the lambdoidal on each side, a short distance above the asterion. The epactal bone shown in Fig. 31 was complicated with a multitude of Wormian bones, many of which, very minute, pertained to the outer table only, and, falling out, left the broad, indefinite border shown

‡ Zeitschrift für Ethnologie, 1888, p. 470.

in the figure. In our series of 88 we have 5 *ossa Incœ* as true and typical as these—a percentage of 5.68.

Fig. 33 represents the only specimen in the Salado series of what Anoutchine calls the incomplete *os Incœ*. In this the left third of the transverse suture is persistent and, connecting with the lambdoidal suture above by an almost vertical suture, separates from the rest of the occipital a triangular bone which probably represents one original point of ossification. This added to the complete *ossa Incœ* gives us 6 specimens or a percentage of 6.81 of both these forms combined.

As an anomaly which is sometimes confounded with the epactal bone, though having a very different embryologic origin, the quadrate bone, or *os quadratum*, is to be considered. Fig. 34 illustrates the only specimen in the collection which can with any propriety be classed under this head, and it is neither a large nor a typical specimen. Some might be inclined to regard it as a triquetral bone. One instance of this form in 88 occiputs gives us a percentage of 1.13.

Immediately above the apex of the quadrate bone in Fig. 34 is seen an open space, which was evidently once filled by a small *os sagittale*.

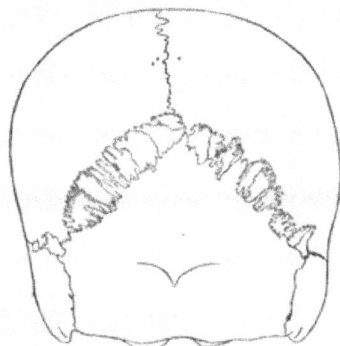

FIG. 35.—Apical bone associated with Wormian bones (No. H. 51).   FIG. 36.—Possible vestige of transverse suture (No. H. 15).

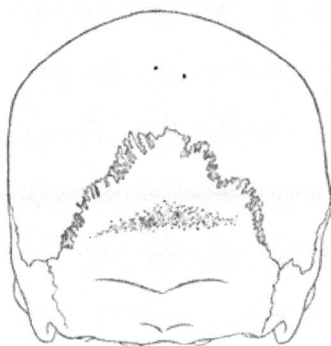

Every separate ossicle or collection of ossicles observed at the apex of the occipital, except a quadrate or an epactal bone, is placed in Virchow's class of *ossa triquetra seu apicis* (apical bones, we shall call them), even when it lies entirely on one side of the median line or is one of the numerous series of Wormian bones like that shown in Fig. 35. It has been found difficult or impossible to draw a definite line of distinction between such and the most typical *os apicis*.

In including all these forms in this class we may have exceeded the limits set for themselves by other observers, and this may account for the large number (16) and the comparatively high percentage (18.1) of *ossa apicis* which this collection furnishes. But if none but the most certain examples were tabulated we would still, probably, have a higher percentage than is to be found in any other race.

There is one anomaly which we have not tabulated, namely, the vestige of the transverse suture which is sometimes seen in the neighborhood of the asterion on one or both sides and which often extends but a few millimeters in the direction of the median line. This is omitted because of the uncertainty attending the examination of minute examples arising from abrasions to the outer table, post-mortem marginal fissures, and other injuries common in these old and friable bones.

In this connection I introduce **Fig. 36**. Here we see a peculiar punctured or honeycombed appearance of the outer table in the line of the transverse suture. It seems to be a vestige of that suture of unusual character confined to the outer table. It is the only specimen of its kind in the Hemenway collection, but there are some similar formations in the general collection of our museum.

The following table is made up of four of Anoutchine's tables, consolidated, with the following modifications: (1) All the races are placed in one order and are called by one name. (2) A title in his table of "Americans in general" is omitted; it would serve in connection with this paper to confuse rather than to enlighten the reader; its figures are obtained merely by adding those of the "Peruvians" to those of "Americans not Peruvians." (3) The Saladoans have been added and placed at the head of the list. Anoutchine's percentages are based on a liberal number of specimens, ranging from 157 in Australians and Tasmanians to 6,871 in Caucasians, in general. The Peruvian specimens are 664, and the Americans (not Peruvians) are 390 in number.

TABLE A.—*Showing the percentage of the Inca bone and allied formations as found in various races.*

Races.	Complete os Incæ.	Complete and incomplete os Incæ.	Os quadratum.	Os triquetrum seu apicis.
Saladoans	5.68	6.81	1.13	18.1
Peruvians	5.46	6.08	1.05	16.5
Americans, not Peruvians	1.30	3.86	0.26	5.63
Negroes	1.53	2.05	2.11	1.19
Malays and Polynesians	1.09	1.42	0.76	0.43
Mongolians	0.56	2.26	0.57	3.62
Papuans	0.57			
Caucasians in general	0.46	1.43	0.18†	1.50
Caucasians of Asia	0.51	1.70	0.41	2.36
Europeans	0.45	1.09	0.13†	1.42
Melanesians		1.65	6.92	2.87
Australians and Tasmanians	0.97	0.61†	0.64	0.61†

The above table speaks for itself and but little comment is necessary. It shows a most remarkable correspondence in the frequency of these anomalies between the Saladoan and Peruvian races. It shows also that, while in respect to three of the anomalies the Peruvians are widely separated from the rest of the human race, as heretofore studied, the Saladoans are still farther removed. In short, they out-Inca the Incas.

It has been maintained[*] that the artificial pressure to which Peruvian skulls were subjected produced the anomaly of the epactal bone. We consider that the arguments in favor of this theory are already successfully refuted, but will nevertheless add to the refutation such testimony as the Hemenway collection offers. The Saladoan skulls bear not the slightest evidence of intentional depression or distortion of any kind, especially of that sort produced by the application to the forehead of the head board, such as the Peruvians once used and some Indians of the northwest coast still use. A certain amount of accidental or unintentional occipital depression is to be found in the majority of the skulls, due apparently to the use of a wooden-backed baby basket with an insufficient pillow; but it is a depression of no greater degree or frequency than is found in many American races among which the epactal bone is comparatively rare. Furthermore, it is not in the most depressed occiputs of the Saladoan skulls that the epactal bone is most common, but in those that are fairly rounded and prominent.

§ 19. FACIAL INDICES.

Being somewhat uncertain as to the true position of the ophryon in these skulls, we took neither the ophryo-mental nor the ophryo-alveolar measurements of Broca, and hence we were unable to compute the facial indices of that author. We have contented ourselves with securing the facial heights of the Frankfurt agreement, which have the definite point of the nasion for their upper landmark, and from these we have computed four indices prescribed by the agreement, namely: The total facial index of Virchow, the total facial index of Kollmann, the upper facial index of Virchow, and the upper facial index of Kollmann. (Tables XXVI to XXXIII, inclusive.)

As much as we have gained in precision by this selection we have lost in another way, since

---

[*] Dissertation sur les races qui composaient l'ancienne population du Pérou. Par M. L.-A. Gosse, Docteur en Médecine. Mémoires de la Société d'anthropologie, vol. I.

the data for comparison of the French measurements are rich, while those for the measurements of the German school are meager. Scattered through the pages of the *Zeitschrift für Ethnologie* and the accompanying *Verhandlung der Berliner Gesellschaft für Anthropologie, Ethnologie und Urgeschichte* there are many separate papers by Prof. Virchow (see Table LXXIX). From these we have prepared, with the expenditure of considerable time and care, Table No. LXXX, and from this we quote a few items for comparison in the facial indices.

We have compiled the following eight tables of comparison, which appear in this section, largely from our own very insufficient special series of 101 before referred to (Table LXXXI). Tables B, D, F, H, show the relations of this Saladoan collection to various races of the world, and Tables C, E, G, I, show its relations to other Indian tribes. Where anything is added from Table No. LXXX the source is indicated in a footnote.

The facial index of Virchow, which is the product of the naso-mental height multiplied by 100 and divided by the facial width of Virchow—a line uniting the inferior extremities of the malo-maxillary sutures—has been computed in 19 skulls. These indices are shown in Tables XXVI, XXVII, where it appears that they vary from 102.85 to 131.25, and that their average is 117.64. For this index, according to the "agreement," the dividing point between the two classes of broad faces and narrow faces is an index of 90, all above this being narrow and all below, broad. These skulls are therefore decidedly narrow faced, but so it would appear are all races as represented in our series of 101, as shown in the following tables:

TABLE B.—*Facial index of Virchow among various races.*

Races.	Number of skulls.	Average index.
Europeans	5	131.77
Negroes	3	127.83
Chinese	2	118.78
Fiji Islanders	2	118.61
Eskimos	6	118.35
Saladoans	19	117.64
Australians	2	117.02
Japanese	2	116.61
Sandwich Islanders	2	115.71
North Americans, Indians excluding Saladoans	30	114.83
Malays (Virchow)	3	*111.59

*From Table LXXX.

TABLE C.—*Facial index of Virchow among American tribes.*

Races.	Number of skulls.	Average index.
Pawnees	1	123.15
Pah Utes	5	117.72
Saladoans	19	117.64
Sioux	4	116.01
Californians	8	115.91
Apaches	4	114.72
Chippewas	2	113.63
Navajos	3	110.29
Poncas	3	108.31

We have been able to compute the upper facial index of Virchow in 34 skulls. This index is the product of the naso-alveolar height multiplied by 100 and divided by the facial width of Virchow. In the tables of this measurement (Tables XXVIII, XXIX) we find that the minimum is 62.22, the maximum 79.59, and the average 69.82. For this index 50 constitutes the point of division between broad and narrow upper faces. The skulls in this group, then, are all distinctly of the latter class. In the tables below it will be seen that there are no averages below 50. In other words, there are no broad upper faces in our special series of 101.

TABLE D.—*Upper facial index of Virchow among various races.*

Races.	Number of skulls	Average index.
Europeans	14	75. 82
Negroes	6	74. 35
Chuckchees	2	73. 48
Eskimos	11	72. 90
Fiji Islanders	2	72. 48
Chinese	2	72. 02
Australians	5	71. 49
North American Indians, excepting Saladoans	43	70. 70
Botocudos	11	*70. 00
Saladoans	34	69. 82
Japanese	2	68. 94
Sandwich Islanders	6	68. 19
Various Malaysians	24	*67. 90
New Zealanders	4	66. 85
Motilo	1	*66. 00
Yucatecs	1	*65. 70

* See Table LXXX.

TABLE E.—*Upper facial index of Virchow among American tribes.*

Races.	Number of skulls	Average index.
Seminoles	2	81. 17
Minnetarees	2	79. 36
Pah Utes	7	72. 49
Sioux	4	71. 99
Californians	10	70. 95
Pawnees	2	70. 66
Saladoans	34	69. 82
Apaches	6	69. 01
Ponkas	4	68. 51
Navajoes	4	66. 88
Chippewas	2	65. 94
Cheyenne	1	65. 13

In 17 skulls we have been able to ascertain the total facial index of Kollmann, which is found by multiplying the naso-mental height by 100 and dividing the product by the bi-jugal width. The tables of this index (XXX, XXXI,) present a minimum of 81.53, a maximum of 97.65, and an average of 88.01 The classes of this index, like that of the facial index of Virchow, have their dividing point at 90; all skulls with an index below that being chamæprosopic or low-faced, and all above that being leptoprosopic or high-faced, the equivalent of Virchow's narrow-faced skulls. Our Saladoan skulls, therefore, which, according to the classification of the Virchow index, are all narrow, are, according to the classification of the Kollmann index, mostly broad (low) and have a slightly broad average.

In the following table (F) of ten different races, in which only two races—Europeans and Negroes—have high faces, the Saladoans appear in a median position and nearer the true Mongolians than to other Indian tribes:

TABLE F.—*Facial index of Kollmann among various races.*

Races.	No. of skulls.	Average index.
Negroes of Africa	6	*95. 49
Europeans	5	92. 80
Negroes of America	3	91. 95
Fiji Islander	1	89. 58
Botocudos of Brazil	3	*89. 10
Japanese	2	88. 95
Chinese	2	88. 63
Saladoans	17	88. 01
Eskimos	6	87. 71
Sandwich Islanders	2	84. 86
Australians	2	84. 59
Various Malaysians	8	*84. 30
North American Indians (excluding Saladoans)	29	*83. 74
Goajiros of Venezuela	8	*83. 30

* See Table LXXX.

In the following table (G) of nine North American tribes it would appear that the Saladoans have higher faces than any other tribe:

TABLE G.—*Upper facial index of Kollmann among various tribes.*

Races.	Number of skulls.	Average index.
Saladoans	17	88.01
The Rock Bluff skull	1	*86.50
Californians	8	85.66
Apaches	4	84.87
Pawnees	1	84.78
Pah Utes	5	83.48
Navajoes	3	83.22
Sioux	4	82.99
Poghas	3	82.74
Chippewas	1	80.99
The Calaveras skull	1	*76.99

*From Table LXXX.

In 27 skulls the upper facial index of Kollmann has been computed (Tables XXXII, XXXIII) its minimum is 47.95, its maximum 60.93, and its average 52.48. In this index, as in the analogous index of Virchow, the highest figure for low or broad upper faces is 50. Of the Saladoans 6 out of 27 belong to this class; the rest have high upper faces, and the average is leptoprosopic. As shown in Table I, the Saladoans for this index have higher upper faces than other Americans in general; but three American tribes exceed them in this particular.

TABLE H.—*Upper facial index of Kollmann among various races.*

Races.	Number of skulls.	Average index.
Chuckchees	3	55.47
Eskimos	10	54.03
Veddahs	2	*73.80
Chinese	2	53.71
Negroes	6	53.22
Europeans	11	53.06
Fiji Islanders	1	52.77
Botocudos	1	*52.60
Japanese	2	52.58
Saladoans	27	52.48
North American Indians, excluding Saladoans	42	51.09
Sandwich Islanders	3	50.35
Australians	3	50.23
Various Malaysians		48.60
New Zealanders	1	48.54

*From Table LXXX.

TABLE I.—*Upper facial index of Kollmann among American tribes.*

Races.	Number of skulls.	Average index.
Seminole	1	58.33
Minnetarees	2	58.05
Californians	10	52.62
Saladoans	27	52.38
Pah Utes	7	52.03
Pawnees	2	51.29
Sioux	4	51.18
Navajoes	4	53.90
Ponkas	4	50.61
Apaches	6	49.55
Cheyennes	1	47.65
Chippewas	1	47.49
The Rock Bluff skull	1	*47.00
The Calaveras skull	1	*42.60

*From Table LXXX

The upper facial index of Virchow is sometimes called "*Oberkieferindex*," an excellent name, for it is indeed an index of height and width of the superior maxillary bone. Now if this shows narrow indices, and Kollmann's method shows broad indices, it is evident that the cause is **consid**erable lateral development of the malar bones.

§ 20. GERMAN PROFILE ANGLE.

In 44 skulls we have determined the German profile angle or *Profilwinkel* of the Frankfurt agreement. We place these angles on record (tables XXXIV, XXXV) more for the advantage of future students than for any benefit they may be to us in the comparative study of this collection, since among the craniometrical literature to which we have access we find few data on this point. The following table we have compiled from works of Wieger[*] and Tarenetzky,[†] including in its proper order the average of the Salado series:

TABLE J.—*German profile angle in various races.*

Race.	Number of skulls.	Average index.
Russians (Tarenetzky)	184	87.70
Americans (Wieger)	15	85.21
Peruvians (included in Americans)	3	84.33
Europeans (Wieger)	15	84.13
Saladoans	44	83.25
Egyptians (Wieger)	19	82.31
Negroes (Wieger)	16	80.52

These figures are somewhat contradictory to those of the gnathic index of Flower, which is designed to express the same character, still the relation of the Saladoans to the Europeans is much the same according to both systems of measurement.

The following is a table of profile angles taken from skulls of various races and tribes in the Army Medical Museum and arranged in order from the highest to the lowest.

TABLE K.—*German profile angle among various races, Army Medical Museum.*

Races.	Number of skulls.	Average index.
Europeans	12	86.25
Mongolians	6	84.75
Sandwich Islanders	8	83.66
North American Indians (exclusive of Saladoans)	43	83.51
Saladoans	44	83.25
Eskimos	10	82.60
Fiji Islanders	2	82.00
New Zealanders	3	81.50
Negroes	5	80.10
Australians	2	78.75

The "North American Indians" grouped under one head in the above table are divided into their separate tribes in the following table, the Saladoans being included in their proper order.

TABLE L.—*German profile angle among American tribes.*

Tribes.	Number of skulls.	Average index.
Minnetarees	2	86.00
Ponkas	4	86.00
Pawnees	2	85.75
Seminoles	2	85.50
Cheyennes	1	85.50
Navajoes	4	83.25
Sioux	4	85.25
Chippewas	3	84.25
Apaches	6	83.58
Pah Utes	6	83.25
Saladoans	44	83.25
Californians	10	79.55

[*] Die anthropologische Sammlung des anatomischen Instituts der Universität Breslau, bearbeitet von Dr. G. WIEGER, in Archiv für Anthropologie, Vol. XV, Supplement, 1885.
[†] Review of article by A. TARENETZKY, in Archiv für Anthropologie, Vol. XVI, 1886.

NOTE ON THE MANNER OF TAKING THE GERMAN PROFILE ANGLE.

This measurement is taken by Spengel's craniometer, an instrument of great accuracy but of rather limited usefulness. A close description of its mechanism is too long to be given here; for such we refer to *Harless's Lehrbuch der Plastischen Anatomie*, zweite Auflage, Stuttgart, 1876, pp. 506 *et seq.*

It is sufficient for our purpose to say that, as regards the facial angle, the craniometer consists of a strong metal table whereon the skull is placed vertex downwards with its right side facing toward the operator and adjusted in the plane of the Frankfurt agreement; and of a goniometer in a plane vertical to that of the table.

It is not often that the skull is sufficiently symmetrical to allow the four points, two supra-auricular and two suborbital, of the required plane to be placed in the same level. It is practically impossible in cases where this may be done to then find the *points de repère* of the sagittal plane vertically one above another. As the goniometer is vertical to the table which serves as a fixed point from which to determine the desired plane, it is evident that in order to use it, the points in the sagittal plane must be vertically disposed. Therefore we place the skull so that the alveolar point is exactly above the nasion while both are on the midline of the machine and face the goniometer. Care is taken to see that some points in the posterior part of the sagittal plane are also in the midline. The skull is then so adjusted that the supra-auricular and suborbital points of the right side, which, as stated, faces the operator, are in the same horizontal plane. The goniometer is put in position and the angle is read.

To sum up: The angle given in this report is, except in cases of skulls with the right side broken away, taken with the skull in such a position that the sagittal plane is vertical and the right side of the Frankfurt plane is horizontal.

§ 21. GNATHIC INDEX.

In 39 cases we have been able to calculate the gnathic index of Busk and Flower, which is found by multiplying the length of the basilo-alveolar radius by 100 and dividing the product by the length of the basilo-nasal radius. The results are shown in Tables XXXVI, XXXVII, and XXXVIII, in which we find (according to Flower's classification) but two skulls that are prognathous (above 103). There are 10 mesognathous (98 to 103) and the remaining 27 are orthognathous (98 and below). The minimum of the series is 88.78 and the maximum 110.11. The average, 95.92, is orthognathous to a high degree and allows us, in respect to the character expressed by the gnathic index, to class this people along with the highest European races.[*]

Gosse states that one of the effects of the occipital deformation, such as these skulls exhibit (*tête déprimée par derrière*), is to diminish the projection of the lower part of the face.[†] Possibly we may thus explain the marked orthognathism of the Saladoans. Nevertheless we fail to discover any direct relation between the facial angle and the occipital contour in this group. Exceedingly flattened occiputs may be found as often among skulls having high as among those having low indices, and the average index of the apparently normal skulls (94.10) is less than that of the obviously flattened, when, as an inference from Gosse's proposition, we might expect it to be higher.

§ 22. ALVEOLO-SUBNASAL PROGNATHISM.

The important character of alveolo-subnasal prognathism we have examined in 27 skulls, according to the rules established by Topinard,[‡] and we have tabulated the angle and the index of this prognathism with the vertical and horizontal measurements which constitute the factors of the latter. (Tables XXXIX–XLII.)

Skull II. 43 has the lowest index, 14.28, and the greatest angle, 82°. Skull II. 57 (Plate L) has the highest index, 61.53, and the smallest angle, 59½°.

The average index of the series is 37.27° and the average angle 70.03°. In the tables given by Topinard§ Americans are not included. His average index of the Malays, 37.42, is nearest to that of the Saladoans, and the factors of the index are much the same, the horizontal being 6.5 in both races. The Malay angle, 69.7°, though not the nearest to that of our collection, is but little removed from it. The angle of the Polynesians, 70.8°, and the angle of the Indo-Chinese, 70.1°, are nearest to that of our collection. Angles of other Mongoloid races, 72.6 to 74.0, are slightly higher, and consequently may be supposed to indicate some evolutionary advancement. His highest average Caucasian angle of 81.8 is not as high as the highest Saladoan, and his lowest average Namaquois of 58.2 is lower than the Saladoan lowest.

---

[*] Topinard: op. cit., p. 94.
[†] Essai sur les déformations artificielles du crâne, Paris, 1855, p. 68.
[‡] "Du prognathisme alveolo-sous-nasal." Revue d'anthropologie. Paris, Vol. 1, 1872, p. 642 et seq.
§ *Op. cit.*, p. 668, and Éléments d'anthropologie, p. 888.

§ 23. THE ORBITAL APERTURE.

Orbital apertures to the number of 38 have been measured according to Broca's instructions, and the indices computed. (See Tables XLIII, XLIV.) Of this number but 2 come within the limit of Broca's class of microseme, or orbit with a low index (below 83.0). These are in skull H. 1 with an index of 82.02, and in skull H. 22 with an index of 81.81. There are 11 in the class of mesoseme, or orbits with median indices (88.9 to 83.0) ranging from 86.61 in skull H. 6 to 88.75, in skull H. 13. The remaining 25 are megaseme, having high indices (89.0 and above). One orbit skull H. 36, is as high as it is broad, having an index of 100, which is the maximum of this group.

In his monograph on the orbital index Broca gives average indices for 66 tribes and divisions of the human family.* Twenty-six of these are megaseme, and to this class all the American races which he mentions, 15 in number, belong. Here, too, belongs our group with its average index of 91.10.

The people having an average index nearest to that of our group are the Indians of our Northwest coast (91.12), while the flatheaded Peruvians (91.50), ancient Yucatees (91.41), modern Mexicans (90.82), Patagonians (90.81), and North American Indians in general (90.75) are not far removed.

There are some items in the table of Broca which seem to show that antero-posterior deformation of the skulls tends to decrease the orbital index. Thus in nondeformed Peruvian skulls the index is 92.20, while in the deformed it is 91.50, and in ancient Mexican skulls the nondeformed have an index of 93.12, the deformed an index of 90.02. These are instances of deformity from intentional frontal pressure (deformation relevée). From the testimony of our collection it does not appear that the accidental occipital pressure has any effect. Of the 38 skulls whose orbital measurements are recorded in the tables (XLIII, XLIV) 11 belong to the apparently normal group. The average index of the latter is 91.06, which agrees closely with that of the rest of the group.

§ 24. NASAL CHARACTERS.

Nasal index.—Forty-four skulls were in a sufficient state of preservation to allow the measurements of the nasal orifice to be taken. As will be seen by the accompanying tables (XLV, XLVI) the average is 51.66, which would place them in the mesorrhinian division of Broca, i. e., where representatives of the Mongoloid races usually stand. The variation in this index is wide, however, extending from leptorrhinian to extreme platyrrhinian.

Inferior border of nasal aperture.—The inferior border of the nasal aperture, échancrure, is of a pretty high type, to judge from the meager statistics of other races to which we have access. Topinard in his *Éléments d'anthropologie*, gives six standards of comparison or classes for this feature as follows: A, the sharp border; A', the slightly rounded border; B, the thick rounded border; C, the border divided into two lips or sometimes three or level (plate-forme); D, the depressed border, first stage of the simian groove; E, the simian groove. These six variants are named in the order of their supposed morphological advancement, A being the highest and E the lowest. Elsewhere ‡ in a monograph older than his last text-book he recognizes but five types, as he had not then apparently made a distinction between A and A'. Hence, in the comparisons which follow these forms are given both separately and combined. In our collection we find so many grades of difference between these standards that it is often difficult to assign a specimen to one or the other; our decisions are often arbitrary, still we do not think we could improve the classification if we would and in all doubtful cases we have decided with special care.

In the Salado series among 48 nares in which the inferior borders can be studied we find them divided as follows: Class A, 15; Class A', 13; Class B, 8; Class C, 6; Class D, 5; Class E, 1. The statistics given by Topinard are in numbers only. We have computed them in percentages (as we have also computed those of the Salado series), in order that we might more easily make com-

---

* Recherche sur l'indice orbitaire, Revue d'anthropologie, Vol. IV, 1875, pp. 616, 617.
† P. 300 et seq.
‡ De prognathisme alcéolo-sous-nasal, in Revue d'anthr., 1872, pp. 634-639. Du bord inférieur des narines sur le crâne et des caractères de supériorité et d'infériorité qu'il fournit, in Bull. Soc. anthr., 1881, pp. 184-192.

parisons, notwithstanding that Topinard's series are very small, he reports on only twelve Mongolian skulls and his highest series, New Caledonians, numbers only 74.

The following table (M) is based on one in Topinard's *Éléments d'anthropologie générale*, p. 802.

TABLE M.—*Inferior border of nasal aperture in three races.*

Class.	Auvergnates.	Saladoans.	New Caledonians
A	52. 65	31. 25	....
A′	20. 54	27. 68	2. 70
A + A′	72. 59	58. 83	2. 70
B	9. 58	16. 67	2. 70
C	13. 69	12. 50	40. 54
D	4. 16	10. 42	33. 78
E	0. 00	2. 08	20. 27

The following (Table N) are percentages placed in the order of numerical importance of classes A + A′ as they occur in various races. They are computed from figures given by Topinard[*], Saladoans added.

TABLE N.—*Inferior border of nasal aperture, Classes A + A′.*

Race.	Percentage.	Race.	Percentage.
Lower Bretons	83. 87	Malays	11. 90
Auvergnians	72. 59	Nubians	4. 55
Saladoans	58. 33	Other Africans	0. 00
Mongolians	33. 33		

From the foregoing it would appear that the Saladoans come next to the Europeans, in the prevalence of a high form of the feature under discussion, and that they are farther above the Mongolians than they are below the Auvergnians.

Position of septum.—In 28 skulls in which the *septum narium* is preserved we find that it is straight in 4,[†] deflected to the left in 11,[‡] and deflected to the right in 13.[§]

Anterior nasal spine.—We find cause for dissatisfaction in applying Broca's instructions[||] to the description of this feature in the present series. We often encounter a long, sharp ridge extending from the extremity of the spine downwards to the alveolar point; this ridge renders spines which are very prominent when viewed from above or below quite subdued when viewed laterally, according to Broca's instructions. Thus, if it were not for the existence of such a ridge, the spine of skull H. 8 would belong to class 5 of Broca, whereas with this ridge it must be placed in class 1 (see Plate 8); but, as we have no other system of description than that of Broca, we have employed it here.

Of 43 well-preserved anterior nasal spines we have 9 of class No. 1 or the least salient, 20 of class No. 2, 12 of class No. 3, 1 of class No. 4, and but 1 of class No. 5 or the most salient. See Table XLVII.

Nasal synostosis.—In seventeen of these skulls out of forty-two examined there is synostosis of at least the upper part of the internasal suture. The percentage then of nasal synostosis in some degree is 40.5. We refer to the upper end of the suture more particularly, because the lower parts of the bones are often broken away. A partial synostosis of the suture at its lower end should not be reckoned in with the others, as it may be the result of some traumatism. Skull H. 30 is probably a case of this kind, as its nasal index is very low and its nasal bones deflected. It is not counted in reckoning the percentage.

---

* *Op. cit.* pp. 801, 802.
† Nos. H. 1, H. 4, H. 22, and H. 29.
‡ Nos. H. 7, H. 18, H. 19, H. 21, H. 23, H. 27, H. 30, H. 32, H. 34, H. 41, and H. 45.
§ Nos. H. 3, H. 6, H. 8, H. 10, H. 11, H. 16, H. 17, H. 20, H. 35, H. 40, H. 43, H. 44, and H. 56.
|| BROCA: Instructions craniologiques et craniométriques; Paris, 1875; Planche VI.

As it is not reasonable in the present state of our knowledge to regard nasal synostosis as possible in children, we disregard four of their skulls, not letting them affect the figures either way.

It is to be noted that there is a partial synostosis in H. 17, a young skull with the basilar suture open and third molars uncut. This is the skull which is so very notable for showing utter disappearance of the sagittal suture.

§ 25.—THE PALATE.

While we have taken four measurements of the palate and one palato-alveolar measurement we have computed only one index, that of Virchow, which depends on the palatal length, from the inner alveolar border between the incisors to the point of the posterior nasal spine, and on the palatal width, taken at the level of the second molars. This we find to be essentially a maximum width, and we prefer in this case the directions of the Frankfurt agreement to those of Topinard as being the more exact. The index is computed by multiplying the width by 100 and dividing the product by the length.

In 32 skulls whose palatine indices we have been able to compute (Tables XLVIII, XLIX) the minimum index is 62.74—which indicates a very long palate—the maximum 84.61, and the average 72.94. Only 3 indices exceed 80, and, therefore, 29 out of 32 are leptostaphylin or long-palate. As none reach the figure 85 the remaining 3 are mesostaphylin or median-palate, while none are brachystaphylin or short-palate.

This series may be said to throw no light on the question of the relationship between the palatine and cephalic indices. It has been shown that in some races a long palate goes with a long skull. In the Saladoans we have a long palate associated with a short skull; but if we admit that the skulls are shortened by artificial means applied to the brain-case, only, we must consider even this negative evidence worthless.

With regard to a correspondence between the face and the palate our series offers better testimony. All the faces, as expressed by their indices, are long; so also are all the palates.

Not only is there this general agreement, but there is to a certain extent an individual agreement in this respect. In order to elucidate this point we have prepared a table (O) given below, in which we have selected for comparison with the palatine index the upper facial index of Virchow for the reason that its table gives a larger number of examples than that of any other facial index.

In columns 1 and 4 of Table O, the number of the skulls are arranged according to the ordination of the facial index but inversely, i. e., the skull having the longest face comes first, and that having the shortest comes last. In columns 3 and 6 we give the order in which each skull would come if arranged according to the length of the palate, for instance: Skull H. 27 has the second longest face and the longest palate, while skull H. 19 has the seventh longest face and the shortest palate.

TABLE O.—*Relation of palatine index to upper facial index of Virchow.*

Place (inverted) in facial index series.	No. of skull.	Place in palatine index series.	Place (inverted) in facial index series.	No. of skull.	Place in palatine index series.
1	H. 11	9	12	H. 2	10
2	H. 27	1	13	H. 28	18
3	H. 43	7	14	H. 4	20
4	H. 29	2	15	H. 5	4
5	H. 10	5	16	H. 17	12
6	H. 8	13	17	H. 7	13
7	H. 19	22	18	H. 33	14
8	H. 1	3	19	H. 16	21
9	H. 11	6	20	H. 41	19
10	H. 40	15	21	H. 50	16
11	H. 45	8	22	H. 29	17

In a glance at the above table it will be seen that the longer palates, whose relative position is expressed by one figure, belong mostly to the first half of the series of 22, while those having the shorter palates belong to the second half of the series. The sums of columns 3 and 6 show this in another way. The sum of the numbers of ordination of the higher-faced half of the series is little more than half the sum of the analogous numbers of the lower-faced half, the proportion being 56.17 : 100. The most aberrant palate in the first half is that of skull H. 19; the most aberrant in the second half is that of skull H. 5.

A list of palatine depths is given in Table L.

§ 26. THE TEETH.

Dr. G. V. Black in the introduction to his article on "Dental Caries"[*] observes that "caries of the teeth has been known in all historic ages of the world, and wherever prehistoric human remains have been discovered traces of this disease have been found. It seems to be and to have been universal in the sense of affecting all nations and tribes of the human race. * * * It has been thought that the savage races were not so much afflicted as the civilized, but my own study of the remains of ancient peoples will not bear out this opinion. This research has, however, been limited within comparatively narrow bounds—too narrow, perhaps, to serve as the basis of conclusions. Unfortunately the literature of the subject furnishes no data that are of much value in this direction, but what there are strongly support the statements made above. * * * The studies I have been able to make in this direction indicate that the races of men that have eaten largely of acid fruits have had less decay of the teeth than those who have been debarred by their position or climate from the use of such articles of food. Generally those tribes that have subsisted largely upon flesh and grain have suffered more from caries than those that have had a more exclusively vegetable or fruit diet. Our knowledge upon this point is, however, too meager to warrant any lengthy discussion of it."

In the following study of the teeth of the ancient inhabitants of the Salado Valley we have taken occasion to make accurate notes not only of caries but also of all deformities of the dental arch, as well as the tuberculation of the superior molars. The materials afforded are fairly abundant and quite sufficient to institute an extended comparison in these respects with other races, with whose remains the Army Medical Museum is so well provided. Unfortunately the materials illustrative of those races whose diet consists exclusively of vegetables and fruits are not abundant in our collections, and it has been deemed best to limit the comparisons to peoples subsisting almost wholly upon flesh or upon a more mixed class of food. The series selected for this purpose are as follows: A series of the Alaskan Indians, whose dietetic habits are well known and who afford an excellent example of an almost exclusively carnivorous race; an unusually large series of ancient dwellers of the Pacific coast region in the vicinity of Santa Barbara, whose food was, in all probability, of a somewhat mixed character; a good series of skulls of Sioux, who furnish a typical example of the carnivorous tribes of the plains; a series of the so-called mound builders of the Mississippi valley; and a series of the ancient Peruvians, who lived largely on vegetable food.

It is proper to state in this connection that only individuals at or below middle life have been selected, since in those races where the wear is rapid, owing, perhaps, to grit contained in the food, the pulp cavity is soon exposed, or the nutrition of the tooth is affected and disease is set up which can not be attributed, properly speaking, to premature decay or caries. We have taken as a mark of middle life the bony union or synostosis of the cranial sutures, either the sagittal or coronal, and there can be little doubt that it is usually expressive of an age of forty or fifty years. Accurate comparisons beyond this limit are difficult, if not impossible, and are therefore not attempted.

The Saladoans, so far as we are able to judge, were a sedentary people, who dwelt in cities and subsisted almost wholly upon the products of the soil, which they extensively cultivated. Indian corn, squash, and other vegetable products must have formed the chief article of their diet, although the presence of charred animal remains in the ruins of their cities indicate that flesh was occasionally consumed. That their remains are pre-Columbian, and that their occupancy of the Salado Valley extended over many generations appear to be well-established facts. As explained in our introduction, it has been pretty clearly shown that some of the modern Pueblos are very

---

[*] American system of Dentistry, Philadelphia, 1896, vol. 1, p. 730.

closely allied to them in both their habits and customs. Unfortunately paucity of material for the latter precludes comparison of their dental organs, which there can be little doubt would furnish additional evidence of value.

Caries.—The subject of dental caries among the ancient inhabitants of the Salado Valley forms an interesting study, inasmuch as it furnishes us with an excellent example of the effect of a given kind of food operating for a long period in the production of tooth decay. It should not be forgotten, however, that other influences may have been in a measure responsible for much of this disease. Their skeletons generally show a remarkable prevalence of osseous disease, and if we are to judge of them by their nearest living allies the lowered vitality of the whole race had at this early date already begun to manifest itself.

Out of some 80 or more skulls we have been able to select 35 in which the sutures indicate them to have been at or under the middle period of life. Of this number 18, or about 51 per cent. exhibit caries, which in some instances has resulted in almost complete destruction of the teeth. Among this number there are also 7, or 16 per cent, in which there has been loss of teeth and absorption of the alveoli without any evidence of caries being present. Seeing the remarkable prevalence of this disease it is but fair to presume that the loss of teeth in these 7 cases is also due to decay which would bring the total up to something like 70 per cent. Out of the remaining 10, which show no evidence of caries, 2 were of very young persons, between 9 and 12 years, in whom we could not reasonably expect to find the disease developed. If therefore these should be excluded the percentage would still be further increased. Among those skulls beyond the middle period of life, fully 90 per cent show caries and loss of teeth; but of these we have not attempted accurate comparisons.

Of the ancient Peruvians we have been able to examine a much larger series—66 in all—wherein there was no bony union of either coronal, sagittal, or lambdoid sutures. In many of them, as in all the other series, teeth had been lost after death so that doubtless in some instances—where the skull has been considered in the category of "no caries"—if all the teeth were present, caries would sometimes be found and the percentage would be thus affected. These cases, however, would probably be few and little change would be necessary.

In this series there are some 8 or 10 examples in which teeth have been lost without any evidence of caries existing. It is fair to presume that some of these at least if not all are the results of dental decay. Out of the 66 there are 29, or about 44 per cent, which show caries, and if 8, in which there is loss, be added, we have the percentage brought to 56. It is proper to mention here that in this series at least half of the skulls examined were not accompanied by the lower jaw, which if present would doubtless show caries frequently, where it does not occur, in the upper jaw, and raise the average of dental caries in these people to at least 50 per cent, if not higher. Respecting the food of these people the early chroniclers are very explicit and we can not do better than quote Garcilasso de la Vega, who has described it at considerable length. He says: "The maize was the principal food of the Indians." They also ate vegetables of various species which he enumerates and describes. Of their meat diet he says (Bk. VI):

The common people were in general poor in flocks (except in the Collao where they had plenty), and hence they only ate meat when they received it as a gift from the Curacas, or when, on some great occasion, they killed one of the guinea pigs they bred in their houses, called Cooz. In order to alleviate this general want the Ynca ordered these hunts to take place, and that the flesh should be distributed among all the people. They made dried meat of it, called "chatqui," which kept good until the next hunt; for the Indians were very abstemious and very careful in preserving their dried meat. * * * It would naturally be supposed that as there is so much water there would be plenty of fish; but in reality there is very little. * * * In the great lake of Titicaca, however, there are many fish. * * * There are several kinds of wild bees, but the Indians did not raise them in hives. The bees in temperate and hot climates, enjoying good herbage, make excellent honey, white, clean, and sweet. * * * The Indians value it much not only for eating, but also for several medicinal purposes.

F. de Xeres* tells us:

The coast people eat flesh and fish all raw, and maize boiled and toasted.*

* Quoted from the Spanish historians in HERBERT SPENCER's Descriptive Sociology, Division II, Part 1 B—New York (1874).

We have selected a series of 42 skulls of the so-called Mound Builders of the Mississippi Valley. These have been collected for the most part in Illinois, Kentucky, Tennessee, and Wisconsin. Whether or not they represent a homogeneous race has not been accurately determined; but it appears to be pretty well established that they lived largely upon the products of the soil of which maize formed the chief staple. It is also probable that they subsisted to a certain extent upon fish and game, but it is believed that they were tillers of the soil rather than hunters. An examination of this series reveals 16, or about 38 per cent, in which caries is present. Of the remaining 26, in 4 cases there was ante mortem loss with obliteration of the alveoli which, if due to decay, would increase the percentage to about 47.

Passing now to the California Coast Indians we find a people whose diet probably consisted largely of fish, although it is well known that berries, grass seeds, acorns, and various vegetable substances formed a part of their food. In this series of 38 skulls 5, or over 13 per cent, exhibit dental caries.

Of the dwellers of the open plains we include 31 skulls of the Sioux. As is well known these people have lived for many generations upon an almost purely animal diet. The Buffalo, until recently furnished their chief staple of food, very little vegetable substance being consumed. Among this number but 3, nearly 9 per cent, out of 34, show any caries. These skulls were gathered over twenty years ago while game was still abundant in the Sioux country. Those with carious teeth are all from eastern bands who had, even then, begun to use the food of white people to some extent.

Lastly we come to the Alaskan Indians, who were probably the most exclusively carnivorous people in existence except the Eskimo. Out of 42 skulls examined we have failed to find a single case of caries, although abscess and premature loss of teeth are present in 8 cases. We are inclined to believe that abscess and premature loss of teeth is more due to accident and violence than decay. It has often been noted of these people that the teeth are extensively used as a sort of vise for many operations, and it would not be at all surprising if they sustained occasional injuries leading to the formation of abscess and not infrequent loss.

With this evidence before us it can not said that a meat diet is injurious to the teeth or a vegetable diet especially beneficial.

TABLE P.—*Dental caries among different American peoples.*

Peoples.	Total number of skulls examined.	Number showing caries present.	Number showing absence of caries.	Number showing absence of loss with caries.	Percentage without loss.	Percentage with loss.
Saladoans	35	18	19	7	54.1	71.1
Peruvians	65	29	29	8	43.9	55.9
Mound Builders	42	16	22	4	38.0	47.6
Californians	38	5	33		13.1	
Sioux	34	3	31		8.8	
Alaskans	42		42			

Deformity.—The malposition of the teeth or deformity of the dental arch is of very frequent occurrence in the skulls of the Salado Valley people. Out of 30 skulls it is found to a greater or less extent in 16, making over 53 per cent. If we divide them up into incisor, cuspid, bicuspid, and molar deformities we find that there are nine cases of malposition of the incisors, six in which the cuspids are affected, five of the bicuspids, and three of the molars. There is one interesting case in which the canine of the left side had been displaced outward by the persistence of a milk tooth occupying a position between the lateral incisor and the first bicuspid.

There are many of these cases of deformity associated with caries of the teeth, more especially in those situations favorable to the lodgment of particles of food. Deformity appears to have been a fruitful cause of decay.

Among the Peruvians, out of 65 skulls we are able to find only 7, or nearly 11 per cent, in which there was any deformity of the dental arch. In these skulls the arch is well rounded and the teeth are very regular, resembling in this respect the form of arch displayed by the Alaskans.

Among the Mound Builders, in a series of 41 skulls there are 6, or over 14½ per cent, of which nearly all referred to the incisors.

The series of Californians, 36 in all, exhibit but 4 deformities, or a trifle over 11 per cent.

Among the Sioux there were found 4 deformities of the dental arch in 34 skulls, or over 11½ per cent.

The Alaskan Indians on the other hand display a much higher percentage of deformity; for out of 41 skulls 8 deformities were found, making nearly 20 per cent.

TABLE Q.—*Dental deformity among different American peoples.*

Peoples.	Total number of skulls examined.	Number of skulls showing dental deformity.	Percentage of deformity.
Saladoans	30	16	53.3
Alaskans	41	8	19.5
Mound Builders	41	6	14.6
Sioux	34	4	11.7
Californians	36	4	11.1
Peruvians	65	7	10.9

Tuberculation.—Prof. Cope* has recently called attention to the absence or slight development of the postero-internal tubercle of the second upper molar in certain races. According to his researches the Eskimos generally have but three tubercles upon the grinding surface of the last two superior molars, representing the tritubercular condition, while the Negroes and Malays display four tubercles upon these teeth, which are, therefore, quadritubercular. These differences are marked and very constant in these races and serve to distinguish two extremes of tuberculation. Among the various tribes of American Indians, however, certain intermediate steps are met with, which in the groups considered we have endeavored to represent by percentages.

Upon the first molar there are always four principal tubercles (two external and two internal) and the grinding face of the crown is always square. In the Negro and Malay the second, and not infrequently the third, molars are similarly constituted; but in the Eskimos the second and third molars bear only three principal tubercles, of which two are external and one internal. The internal cusp is large and crescentic in outline and covers the entire internal aspect of the grinding surface; but it sometimes happens that a faint trace of the fourth cusp is present in the form of a slight ledge or cingulum at the postero-internal angle of the crown. Those skulls in which the second molar has its full complement of tubercles we have marked 4; those in which the tooth displays a trace of the fourth cusp we have marked 3½, while those in which there are only three tubercles we have marked 3.

Taking the Alaskans as the extreme of the tri-tubercular type we have in 43 examined skulls 29, or over 67 per cent, in which the second molar bears 3 tubercles; 8 of the 43, or over 18½ per cent, display traces of the fourth cusp, and 6 of the series, or nearly 14 per cent, have the fourth cusp fairly well developed.

Out of a series of 71 skulls of the ancient Californians 44, or nearly 62 per cent, are tritubercular; 15, or about 21 per cent, have traces of the fourth cusp, and 12, or nearly 17 per cent, have all the tubercles developed.

The series showing the next highest percentage of the tri-tubercular type is that of the Mound Builders, in which out of 37 skulls 15, or 40½ per cent, are tri-tubercular; 4, or nearly 11 per cent, have the tubercles 3½, and 18, or over 48½ per cent, have all the tubercles present.

The condition of the second molar in the Saladoan skulls gives the following results: Out of 23 examples 9, or about 39 per cent, are tritubercular, and the remaining 14, or nearly 61 per cent, are more or less quadritubercular.

Next come the Peruvians, in whom 19 out of 53 skulls, or about 36 per cent, are tri-tubercular, 14 or nearly 26½ per cent have the 3½ tubercle, and 20 or over 37½ per cent are quadri-tubercular.

Lastly we come to the Sioux, of whose skulls 37 are represented in this series. In these 6 or

---

*Journal of Morphology, Boston, 1888, 1889, Vol. II. pp. 7, etc.

over 16 per cent are tritubercular, 18 or over 48½ per cent have tubercles 3½, and the remaining or slightly over 35 per cent have four tubercles well developed.

From a careful consideration of the facts here set forth it would seem that the nearest allies of the ancient inhabitants of the Salado Valley, if we judge from the prevalence of dental decay, are the Peruvians upon the one hand, in whom caries was almost as frequent, and the Mound Builders of the Mississippi Valley on the other, who also suffered to a considerable extent from tooth-decay   Whether we are to accept the dental condition described as indicating affinity or whether they are to be regarded as the effects of climate, food, and general habits of life we are not prepared to say; but it is more than probable that they have a certain value as expressing race affinity.

The facts relating to the structure of the teeth themselves are important, and we are disposed to attach more weight to them, so far at least as evidence of affinity is concerned, than to the other two classes combined.   The high percentage of the trituberecular second molar in the Alaskan Indians, 67 per cent, is significant and betokens either much commingling or a very near relationship with Eskimos.   In a like manner the percentage of 62 among the Californians is suggestive of near affinity with the inhabitants of Alaska.   The Mound Builders, Salado Valley people, and Peruvians on the other hand are very closely related in this respect, as is indicated by the percentages 40, 39, and 36, while the Sioux stand considerably apart from the rest with a percentage of only 16.

TABLE B.—*Tuberculation among different American peoples.*

Peoples	Total number of skulls examined.	Number showing 3 tubercles.	Number showing 3½ tubercles.	Number showing 4 tubercles.	Percentage of 3 tubercles.	Percentage of 3½ tubercles.	Percentage of 4 tubercles.
Alaskans	43	29	8	6	67.4	18.6	13.9
Californians	71	44	15	12	61.9	21.1	16.9
Mound Builders	37	15	4	18	40.5	10.8	48.6
Sahedoans	23	9	....	14	39.1	....	60.8
Peruvians	53	19	14	20	35.8	26.4	37.7
Sioux	37	6	18	13	16.2	48.6	35.1

§ 27. THE HYOID BONE.

[By JACOB L. WORTMAN, M. D., Anatomist of the Army Medical Museum.]

The following study of the human hyoid arch has been undertaken with a view to the determination of the more exact value of this series of bones in matters of anthropological research.   The subject has received so little attention at the hands of anatomists, especially from this particular standpoint, that there is little or no literature upon it, and we are as yet in comparative ignorance regarding the conditions and characteristics of this chain of bones, even in the best anatomically known races of mankind.

The history of this undertaking dates from the author's connection with the Hemenway Southwestern Archæological Expedition to the valley of the Salado, Arizona, in 1887, whither he was sent by the United States Army Medical Museum to obtain a full series of skeletons of the ancient dwellers of this region.   While engaged in the collection of this material it was noticed that the body or middle piece of the hyoid bone was almost always free, and that the separate pieces, of which the hyoid arch is made up, seldom united into a single bone, even in the most aged individuals.   The hyoid, as the writer had been accustomed to see it in skeletons of whites and negroes, consisted usually of a single U-shaped bone, especially if the individual had passed the middle point of life; and upon consulting a few standard text-books on human anatomy which had been taken into the field for ready reference it was found that this was regarded as the usual or normal condition.

The attention of Dr. Herman ten Kate, the anthropologist of the expedition, was called to the subject, and together we took accurate note of the probable ages, conditions of bone disease, etc., of all the individuals whose hyoids were secured.   In all there were obtained some 97 specimens of various ages, which are now preserved in the collection of the United States **Army** Medical Museum at Washington.

Upon our return to Washington we searched the literature carefully for any statement that would throw light upon the subject, but were unable to find that anything had been said or written upon the subject other than the general statements contained in works upon human anatomy. We accordingly prepared a paper setting forth the principal facts, which was presented to and read before the Congress of Americanists held in Berlin.

One of the chief difficulties with which we had to contend in discussing the general bearing **and** importance of our discoveries was the lack of materials for comparison. Since then the writer has been actively engaged in collecting materials illustrative of the characteristics of the hyoids in the negroes and whites, and he is now in a position to discuss the subject upon a more accurate basis. The sources of materials have been as follows: From Prof. Thomas Dwight, of the Harvard medical school, the Museum has received a record of 33 cases, of which 4 were black, 28 white, and 1 of mixed Mexican and Indian parentage. These specimens were from individuals ranging from 17 to 82 years of age, and include both sexes. From Prof. Towles, of the University of Virginia, the Museum has received 12 specimens of hyoid bones, all from negroes, with the ages attached. From Prof. Matas, of the Tulane University, New Orleans, there are 17 specimens, of **which 12** are from negroes, 4 from whites, and 1 from a Chinese. From a personal collection there **are** 23 specimens, of which 21 are of colored people and 2 are of whites.

What may be considered as a typical hyoid arch of the higher mammalia is to be found in the dog, Fig. 37, which is taken from Prof. Flower's "Osteology of the Mammalia." We prefer to follow this author in the nomenclature of the several elements composing it, which is essentially that proposed by Prof. Owen many years ago. In this we observe first a central unpaired piece or body, which is denominated the "basihyal;" from the outer extremities of this central piece two long slender rods of bone project backwards over the upper edge of the thyroid cartilage and are called the "thyrohyals" or greater cornua. Near the junction of the thyrohyals with the basihyal are attached the distal pieces of two chains of bones which connect the basihyal piece or body with the temporal bones of the skull. The first piece of this series, counting from the basihyal, is the lesser cornu or "ceratohyal"; the second is the "epihyal," the third is the "stylohyal," and the last piece, which finally joins the skull, **is** that called by Prof. Flower the "tympanohyal."

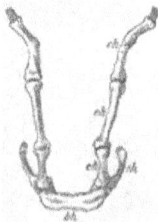

FIG. 37.—Extracranial portion of hyoidean apparatus of a dog, front view; *sh*, stylohyal; *eh*, epihyal; *ch*, ceratohyal (these three constitute the "anterior cornu"); *bh*, basihyal, or "body" of hyoid; *th*, thyrohyal, or "posterior cornu." [After Flower.]

While this might be called the typical **arrangement of** the mammalian hyoid apparatus, it so happens that in many forms, including monkeys and man, the complete bony connection between **the** basihyal and **the** base of the skull does not exist, owing either to the absence **in** this chain of bones of certain elements or their rudimentary condition. In this case a ligament **may** take the place of one or more of these elements, which in human anatomy is known as the stylohyoid ligament.

Prof. Flower, in speaking **of the** human hyoid apparatus, says:[*]

The stylohyal, at first a long styliform piece of cartilage continuous with the tympanohyal, commences to ossify by a separate center before birth, and at a very variable period afterwards is often (but by no means constantly) anchylosed with the tympanohyal and surrounding cranial bones, constituting the so-called "styloid process." This is a condition not met with in any other mammal. Below the stylohyal the greater part of the anterior hyoid arch is represented by a slender ligament (the "styloid" ligament), there being no ossification corresponding to the dog's epihyal.

This view has been generally accepted and it is now commonly taught that the epihyal element of the dog is missing in the human hyoid arch.

A different conclusion upon this important point has been reached by Thomas (de Tours),[†] who, in speaking of the human hyoid arch, says:

The body is the strongest piece of the entire apparatus. This is an osseous lamina curved in the form of an arc; its anterior face, very irregular, is convex from side to side and from above downwards, and is composed of two ob-

---

[*] Osteology of the Mammalia, p. 159.   [†] Éléments d'ostéologie, Paris, 1865, p. 219, Pl. x.

lique planes, the one anterior and inferior, and the other superior. The angle at which these units forms a prominent ridge directed transversely in the sense of the greatest dimension of the bone; the posterior face is profoundly excavated. At each extremity it is articulated with the thyroid cornua. These are two straight bony pieces directed from before backwards and laterally flattened; their posterior extremities give attachment to the thyro-hyoid ligament. These pieces unite, forming an arc, to the circumference of which the larynx is suspended, followed by the trachea and lungs.

Each hyoid chain is composed of three pieces, as in the preceding animals (dog and sheep); the first or superior piece has the form of a very elongated cone with its base above and its summit below, its greatest dimension being three centimeters. Its base articulated with the hyoid prolongation gives it a varying length. Its union with this prolongation takes place at different periods, sometimes at thirty years, sometimes at sixty years. This union is to be always recognized by its nodular appearance, more or less distinct.

The second or intermediate piece has nearly the same form as the first, except that it is much more slender; its length is about two centimeters; its base articulating with the summit of the first piece at the middle of the ligament gives it a very variable length. From its summit proceeds the stylo-hyoid ligament, which terminates in the third piece or small cornu of the hyoid and forms a very acute angle with the greater or thyroid cornu.

This third piece has very often the form and size of a grain of barley, but sometimes it is elongated and styliform, like the intermediate piece. It joins the extremity of the body of the hyoid in such a manner as to form an articulation common to it and the greater cornu. The stylo-hyoid ligament is composed of whitish glistening fibers possessing great elasticity. It is very slender, tapering in its superior and swelling out in its inferior part, which is attached to the small hyoid cornu.

In the normal condition in man the superior piece of the hyoid chain is united by one extremity to the hyoid prolongation and by the other to the intermediate piece. One then finds the styloid process of authors. This osseous stem, 4 or 5 centimeters long, knotted and sometimes curved and twisted, ends in a point, and in certain subjects descends to the angle of the jaw.

It is this disposition, the union of the two superior pieces between themselves and with the hyoid prolongation to form the styloid process, and the other part, the great distance between the preceding piece, and the third part, a distance traversed by the stylo-hyoid ligament, which has caused the error of anthropotomists and has led them to divide the hyoidean chain into two parts—the one which has been described with the hyoid, viz, the small cornu or third piece, and the other which has been attributed to the temporal, viz, the styloid process. They might have easily avoided this error by studying comparatively the hyoid apparatus of man and animals. They might have recognized that the styloid process of man represents the stem formed in the ruminant and in the carnivore by the first two pieces of the chain, and that in man the articulation at a long distance of the summit of the styloid process with the lesser cornu corresponds to the disposition of the third much more movable piece, which descends from the rigid rod to suspend the hyoid in animals.

Several authors in works on human anatomy mention the condition described by Thomas: Meckel, in speaking of the temporal bone, says[*]

The muscular eminences and depressions are, first, the styloid process (processus styloideus), at the posterior extremity of the under edge of the pyramid; this varies much in length and sometimes exceeds two. This process is sometimes entirely free and is often composed of several pieces—a curious analogy with animals.

In Gray's Anatomy it is stated:[†]

The styloid process varies in size and shape and sometimes consists of several pieces united by cartilage.

The writer's experience upon this subject is confined principally to observations upon the adult skull. He has, however, examined a number of foetuses, in which he has always found the styloid process to consist of but a single slender piece of cartilage reaching from the temporal towards the basihyal. It is highly probable that the failure to find the several elements described was due to the age of the specimens examined, all of which were at or before full term.

The most favorable age to select is somewhere between the time when ossification begins and twenty-five or thirty years. Unfortunately, in the average museum specimen of this age the styloid process has not been preserved, and all that one can discover is a short peg of bone wedged in between the two laminae of the vaginal process. In skulls of more advanced age, wherein the several pieces have not only united with the skull but have been joined to each other, it is not an easy matter always to determine the point of union.

In a large series of skulls in the collection of the Army Medical Museum the following is the most common condition: A short distance below or quite at the edge of the vaginal process there is a considerable swelling or nodosity, and if the subject be not too old the remains of a suture are discoverable at this point. Sometimes this nodosity is placed as much as a half an inch below the

* J. F. Meckel, Manual of Descriptive Anatomy (English Translation), London, 1838, Vol. 1, p. 57.
† Gray's Anatomy, 1887, p. 144.

edge of the vaginal process and sometimes quite within its folds. Below this nodosity there can sometimes be seen a second swelling with the same evidences of a suture. Then, again, there are many skulls in which the first nodosity is present, and the process is terminated by a truncated extremity, as if a piece had been attached to it, but had been lost in preparation; and, finally, in some few cases the styloid composed of three distinct pieces was observed, as described by Thomas.

There can be little doubt that the part spoken of by Thomas as the "hyoid prolongation" is the tympanohyal element of Flower, which, there is good reason to believe, is variable in length. There is also little doubt that not uncommonly there is a distinct ossification intervening between the lower end of the true stylohyal element and the ceratohyal piece, or small cornu of the hyoid, which can not be accounted for upon any hypothesis other than that it is the strict homologue of the missing epihyal so constant in the lower forms. It would be a matter of no little interest to determine the frequency of its occurrence in the various races of mankind. (See Figs. 38 and 39.)

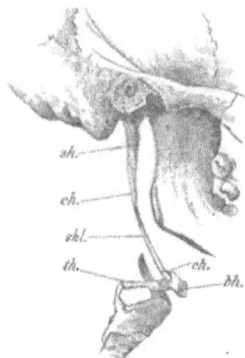

FIG. 38.—Hyoidean apparatus of man. [After Thomas.]

Passing now to the hyoid bone proper, we have to consider the several elements of which it is composed. As is well known, it is generally described in works on human anatomy as consisting of a single U-shaped bone, formed by the union of five pieces. These are known as the body and the greater and lesser cornua. Although there does not appear to be absolute unanimity of opinion among anatomical writers regarding the particular time of life when these elements coössify, we can not do better than to quote here the statements made by the leading anatomical authorities upon this point.

Among the German anatomists Meckel, in his *Manual of Anatomy*, says:

"The hyoid bones, or the hyoid bone, forms an arch which is convex forwards. It is situated behind and below he maxillary, beneath the root of the tongue and the upper part of the neck. It is generally considered a single bone, and is divided into a central portion, or body and four horns, two upon each side; but as these parts remain distinct throughout life it is better to admit five distinct bones, a middle and four lateral. The inferior hyoid bones, or the greater cornua of the hyoid, often vary considerably in form and size upon the different sides in the same subject. They articulate with the central piece by a fibre-cartilaginous mass and sometimes unite in the latter periods of life in one bone."

FIG. 39.—Styloid process of man. [After Thomas.]

Henle, in his *Human Anatomy*, says:

"The great horns of the hyoid bone can also be connected with the body by joint. Many hold this to be the rule."

Hyrtle, in his *Lehrbuch der Anatomie des Menschen*, says, quoting from Meckel:

"The os hyoid is divided into central or body and two lateral cornua, which parts, as they are united by movable articulation or by synchondrosis, and often in old age not coössified, can be considered as so many different or separate hyoids."

Gegenbaur, in his *Lehrbuch der Anatomie des Menschen*, says:

"The great cornua often coössify with the body."

Hartmann, in his *Handbuch der Anatomie des Menschen*, says:

"The five parts of the hyoid bone articulate by movable joint at the small horns and with synchondrosis at the large horns. In old age these parts are ofttimes anchylosed."

Krause, in his *Menschliche Anatomie*, says:

"The great horns are united with the body by capsular ligament, and the joint is an amphiarthrosis. Very often it is only a synchondrosis."

Walter, *Human Osteology*. Berlin, 1798, says:

"It is rare that the entire bone is ossified. It occurs only in very advanced age."

The conclusion which **one** draws from these statements is that the great cornua of the hyoid bone remain free **even in old age in** the majority of examples upon which these observations have been made, and **all** these authorities seem to agree that it **is** only at a very advanced period of life that any of the **hyoidean** elements coössify. Taking **for** granted that the observations of German anatomists **have been** made upon German materials for the most part, one can safely say, if these statements be correct, that this is the normal condition of the German hyoid.

French anatomists make a different statement. Sappey, in his *Traité d'anatomie descriptive*, 1867–72, says:

"At 40 or 50 years, ofttimes before that period, the great cornua are joined to the body. The little horns are also sometimes joined to the body, but only in old age."

Boyer, *Traité d'anatomie*, 1803–9, says:

"With age the great cornua are joined to the body. The small cornua also unite, but this happens much later."

Cruveilhier, *Anatomie descriptive*, 1844, says:

"All the pieces are at first separated by considerable portions of cartilage, afterwards by a very thin layer, which sometimes remains during life."

Portal, *Cours d'anatomie médicale*, 1803, says:

"The borders of the body and the middle of the greater horns ossify first, but they remain epiphyses for a long time, or separated from the body of the bone by a portion not ossified, and which hardens with age. The small cornua remain still longer without ossifying; but in old age not only are all the pieces of the hyoid united, but the stylohyoid ligament is ossified."

Beaunis and Bouchard, Nouveaux *Éléments d'anatomie descriptive*, 1873, say:

"The great cornua are sometimes united to the body by a true movable articulation. The small cornua are habitually movable upon the rest of the bone."

One **would be led to infer from** these statements that the normal condition of the French hyoid, allowing that **the observations of** the French anatomists have been made upon French subjects, is the complete consolidation **of all the five** elements and, if Sappey's statement can be trusted, at a comparatively **early period of life**, so far at least **as** the great cornua are concerned.

It is a difficult matter to **reconcile these statements of the French** and German **anatomists** otherwise than upon the **ground of difference in the structure of the hyoid itself in these two** peoples. It would be interesting **to** determine the truth or falsity of **this supposition.**

English anatomists agree more nearly with the French in their **statements of the hyoidean** pieces. Flower, in his *Osteology of the Mammalia*, 1870, says of the human **hyoid:**

"The thyrohyals or great cornua of the hyoid bone are elongated, nearly straight, and somewhat compressed. They usually become anchylosed before middle life with the outer extremity of the basihyal."

Holden, *Human Osteology*, 1885, says:

"Until the middle period of life the great cornua are united to the body by cartilage, but this ossifies in the progress of age."

H. **Hyde Salter,** in *Todd's Cyclopædia of Anatomy and Physiology*, article, "Tongue," says:

"Ossification begins in the greater cornua; it then takes place in the body, where it begins soon after birth, and finally **in the lesser** cornua, where it does not commence until some time after. It proceeds but slowly, and generally leaves **a thin lamina of** cartilage unossified, so that complete anchylosis into one bone is comparatively rare."

Erasmus **Wilson,** *Human Anatomy*, 1859, says:

"In early age **and in the** adult the cornua are connected with the body by cartilaginous surfaces and ligamentous fibres, but in old **age they become** united by bone."

In Gray's *Anatomy* **it is stated:**

"In youth the cornua are connected to the body by cartilaginous surfaces and held together by ligaments; in middle life the body and greater cornua usually become joined, and in old age all the segments are united together, forming a single bone."

Morton, *Human Anatomy*, 1849, says:

"The cornua are connected to the body by a distinct movable articulation, which generally, however, becomes **anchylosed** later in life."

Just how far the statement of **any of the** preceding authorities is the result of individual knowledge and experience, or to what extent **the** information was drawn from previous authors, or the number of cases upon which the observations were made, does not appear, and for this reason the exact anthropological value of the statements is difficult to estimate. In order to reach the question in a more definite manner we give the results of an examination of 32 specimens of hyoids from whites whose ages are known. For this series the lowest limit in age taken is 35 years, which, although somewhat below the middle point of life, will yet be more nearly comparable to the series of the Saladoans and the negroes which will be referred to later.

The sexes from which the specimens were taken are about equally represented; but the nationality is not given farther than that they were white. Of these 21 show bony union of the greater cornua with the body, and in 11 the cornua are free, giving a percentage of 65 and a fraction for those that are joined. For 24 of the specimens the age given is 45 years and upward, and of these 18 are joined and six are free, making a percentage of 75. A more detailed analysis of the union and non-union is as follows: United upon both sides, 17; united upon the left side, 3; united upon the right side, 1; both cornua free, 11. It may be remarked that in the remaining five specimens under 35 years of age there is one (age 31) which shows union of one of the greater **cornua,** namely, upon the left side. If this was added to the list the percentage would be increased **to 66** and over. However, the percentage of 65 may be regarded as a fair expression of the condition of the hyoid of the white so far as the bony union of the greater cornua is concerned. In those of 45 years and upwards 75 per cent is probably a fair estimate of this condition.

Turning now to the negro, we have altogether a series of 35 hyoids which pertain to persons of 35 years and upward. Of these 27 show bony union of the greater cornua with the body and 8 are free, giving a percentage of 77 and over; 21 are joined upon both sides; 3 are joined upon the left side; 3 are joined upon the right side, and 8 are entirely free. We have previously reported upon a series of 25 negro hyoids,[*] in which the percentage of bony union of the greater cornua was found to be 66. If now we include these 35, we have a series of 60 specimens in which the mean percentage is 70. Of the 35 there are 12 of 45 years and upward, of which 10 are joined and 2 are free, giving 83⅓ per cent. This examination does not take into consideration those cases of mixed blood, since some of the specimens are known to be from mulattoes. Just how this has influenced the percentage is not easy to determine, but it is no more than reasonable to suppose that it has had some effect, and may account in a measure for the close correspondence between the white and the negro in the matter of union of the greater cornua.

In the light of these facts we come lastly to consider the hyoids **of the ancient Saladoans, of** which there are 97 in all, many of them being complete. Some **of this number are not** accompanied by the skeletons to which they belong, owing to the advanced **stage of decay in which** they were found rendering their preservation impossible. In all **cases where the skull could not be** preserved a careful examination was made with a view to the determination of the age from the condition of the teeth, the synostosis of the sutures of the skull, and the angle of the jaw.

We have adopted the system of labeling them Young, Adult, Old, and Very Old. In the category of "Young" we have placed all those specimens under the age of 21 years, or those in which the last molar had not been erupted, the teeth themselves little worn, and the evidence of epiphyses had not yet been obliterated. In the class "Adult" we have placed all examples in which the teeth were fully erupted and all evidence of epiphyses obliterated, but which do not show any bony union of the cranial sutures. In the class "Old" we have placed all those in which the teeth are very considerably worn and the sagittal or coronal suture shows bony union. The class marked "Very Old" we have made to include all those specimens in which the sagittal, coronal, and lambdoidal sutures were synostosed, in which the teeth were entirely gone—their alveoli being absorbed—or were reduced to inconsiderable stubs, and the angle at which the horizontal ramus of the lower jaw joins the perpendicular portion was very open or obtuse. In most of the examples of this class all the sutures of the skull had disappeared, indicating great age.

That part of our material in which the greatest amount of interest centers is, of course, in the classes marked "Old" and "Very Old," and it is more than possible that a certain number of

* American Anthropologist.

anomalies in the premature union of the cranial sutures, as well as the loss of the teeth and the absorption of the alveoli, exist; but we are persuaded to believe that the series is a fairly typical one and exhibits the normal condition of this race in these particulars.

It may be urged that the determination of age upon the basis which we have adopted is not sufficiently accurate for purposes of this kind; but there are few anatomists who would hesitate to pronounce judgment upon the age of a skull from the evidences which we have cited. At all events, we feel that we are entirely within the bounds of reasonable judgment when we say that the classes "Old" and "Very Old" pertain to individuals not under 35 years of age.

Of the class "Very Old" there are 13 examples of the hyoid, in which union of the great cornua with the body is found in 3. In these 3 cases the union is partial, for it is only upon the left side that it exists. It should be stated that in 1 other of these 13 cases the hyoid is represented by one of the great cornua only, so that it is impossible to say whether partial union existed or not upon the opposite side in the case.

Of the class "Old" we have 44 specimens in which bony union of the great cornua with the body of the hyoid exists on both sides in 2, on the left side in 1, and on the right side in 1, making 4 in all. Of these examples 9 are represented by one of the great cornua only, so that it is impossible to say whether partial bony union existed upon the opposite side or not. In all the 4 cases in which partial or complete bony union is found we have discovered skeletal disturbances in the way of exostoses, unusual anchylosis, etc., which would naturally lead to the belief that the union of the hyoid elements was an abnormal condition as well. Be this as it may, however, it will be seen that the percentage of union is very small. Taking both classes in which there are 7 coössi-fications in 57 specimens, we have a percentage of only over 12 as against 65 and 77 of the white and negro, respectively.

This difference is marked, and in our judgment can not be accounted for upon any other hypothesis than that of a natural anatomical distinction which these people possess. In the paper which Dr. ten Kate and the writer prepared upon this material we stated at that time—

That owing to the lack of materials for proper comparison we are unable to make any satisfactory deduction respecting the hyoid at this particular time, and what we here note must be regarded as merely a statement of fact to be correlated in its proper place. * * If, on the other hand, we are to accept the statements of many of the anatomists we have already quoted, then we can say that the very high percentage of free hyoidean elements which we have found in these ancient people distinguishes them markedly from some other races. If, again, it is found that this condition of the hyoid is general in North American Indians, as well perhaps as some other races, it would be interesting to know in what way, if any, it is associated with their language.

These surmises were probably correct, and there appear to be marked distinctions between the hyoidean apparatus of these ancient Saladoans on the one hand and the whites and negroes on the other, a distinction which is indicated by the percentages already set forth.

In a series of 17 specimens recently received from the ancient cemeteries in the vicinity of Zuñi, New Mexico, there are 4 showing bony union of the great cornua and 13 are free. A careful inspection of the skeletons to which they belong gives an indication of age from at least 35 years and upwards. The percentage in this case is 23 and a little over.

From a few specimens (9 in all) of hyoids of the so-called Mound Builders there are 4 coössi-fications, giving a percentage of 44 and a fraction; but this series is too small to be of much value to us.

Regarding the lesser cornua we have not devoted that attention to them that we have given to the greater cornua and body of the hyoid: but if we are to judge from what Prof. Thomas Dwight, of the Harvard Medical School, says, it would seem that they may be entirely absent. In a letter transmitting the record of observations given above, he writes:

The result of the examination of the lesser horns is rather surprising, as it shows that they are very rarely united to the body of the bone, that the mode of connection with the body varies, and that one or both may be entirely wanting. In only one of the 33 hyoids were both lesser horns coössified, and in only 4 others was a single horn thus united. It is generally taught that the joint between the body and lesser horn is synovial. This is certainly true in many cases but not in all. Sometimes the lesser horn is attached by ligament, and at least in one case I have found it held by muscular fibers. In other cases, owing chiefly to the parts having become dry, it was impossible to decide whether this was a true synovial joint or not. In several cases one or both the lesser horns were not found, and it was not always possible to determine whether the absent piece had been lost or had never existed. It was, however, shown beyond question that one or both of these horns may be wanting. One was wanting in a girl of **17 and both** in a man of 55. In a woman, said to be 80, one was wanting and the other probably wanting. In a **man of 37 and** another of 39 one was probably wanting. In a woman 50 and a man 55 both were probably wanting. **When a joint** was found upon the body it was clear that the lesser horns had been lost, which occurred two or three **times; but** the absence of a joint does not show beyond question that the horn was wanting as it may have been held by ligament. It is thought most probable that where the entry has been made " lost or wanting," the bone was originally wanting.

We come now to consider the body of the hyoid bone, and we regret to say that the soft parts, particularly the larynx, could not be included in this study since our material refers almost exclusively to the dry bone itself. The body of the hyoid in monkeys has a distinctive and characteristic form, which according to Flower[*] has a greater vertical than transverse diameter (see figs. 40 and 41). This form of the hyoid body is associated in many of the lower types of monkeys with a membranous sack which occupies the concavity of the bone and protrudes between the lower edge of the body and the upper edge of the thyroid cartilage. It was called the hyothyroidean sac by Cuvier, and the *succus membranaceus* by Wolf. It has an opening at the base of the epiglottis and is said to sometimes communicate with the laryngeal sac which lies just above the vocal chords. According to Eckhard,[‡] this hyothyroidean sac is absent in the anthropoid apes, with the possible exception of the gibbon. We are not sufficiently familiar with the anatomy of the larynx of the anthropoids to state whether any rudiment of this condition is to be found in them; but it would not be at all surprising if this eventually turns out to be the case. We are led to infer that the true significance of the great depth of the body of the hyoid in the monkey is to be explained primarily upon the basis of this sac, whatever its function may be, and that the depth of the body in proportion to its width furnishes an **index of** this distinctively simian feature, which we propose to call the basihyal index.

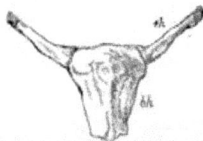

FIG. 40.—Hyoid of baboon; *bh*, basihyal; *th*, thyrohyal. (After Flower.)

FIG. 41.—Hyoid of an American monkey; *th*, thyrohyal; *ch*, ceratohyal; *eh*, epihyal. (After Flower.)

It is therefore with no small amount of interest that we come to examine **this question in the** light of our present material. We have been necessarily compelled to limit our researches to the Negro and Saladoan, for the reason that our materials have proven insufficient **as regards** other races, which are therefore not included. Some difficulty has been experienced in determining just where the measurements should be taken in case the greater cornua are cobossified with the body, which is, as we have seen, the usual condition of the adult Negro **hyoid**. After careful attention to this point we have determined upon the following measurements: The vertical depth is obtained by placing the bone flatwise upon its posterior surface **and measuring** with a pair of calipers or other suitable instrument its greatest diameter in **this direction. The** transverse diameter is taken by placing one arm of the dividers upon the **point of** union of the anterior ridge with the lingual or superior border and measuring to the corresponding point upon the opposite side. In some instances the anterior transverse ridge is not well defined and the point where it terminates **is not** easily made out. In such cases, if there remain any traces of the suture joining the great cornua with the body we measure from this suture where it crosses the superior border to the same point upon the opposite side.

Among the Saladoans the bodies are mostly free and we have had little difficulty in determining the proportion of the depth to the width. In one instance we measured the greatest diameters and found that the proportion of the depth to the width is 52 per cent and a fraction in 45 specimens. In the same series measured between the points indicated above for the transverse diameter the proportion is 54 per cent.

---

[*] FLOWER: Osteology of the Mammalia, p. 140.

[‡] MÜLLER: Archiv für Anatomie und Physiologie, 1847. p. 44.

In a series of 36 negroes the proportion of the depth to the width is 65 per cent and a fraction, or between 11 and 12 per cent more simian. In one case (Fig. 42) the proportion goes as high as 90 per cent, while 75 per cent is not at all unusual in the series.

In the few specimens of the white hyoids which we have the proportion seems to be about 50 per cent, although we have not been able to determine this with any degree of exactness. (See Fig. 43.)

Fig. 42.—Anterior and Posterior views of negro hyoid.

In conclusion we will say that in the present state of our knowledge it is well-nigh impossible to give any intelligent explanation of the facts which have been set forth above, with the possible exception that the greater basihyal index of the Negro is to be accounted for on the basis of his nearer relationship to the monkey. Regarding the coössification of the greater cornua with the body little can be said, but it might be suggested that, since the chief function of the hyoidean appa-

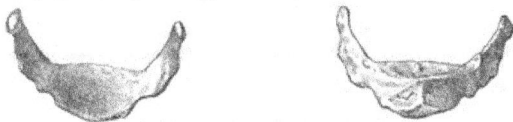

Fig. 43.—Anterior and Posterior views of European hyoid.

ratus is the support of the muscles of the tongue, one would be led to infer that it has something to do with language. It is supposable that in those races where rapid talking and much talking was the rule the hyoidean elements would coössify early, while among those people who speak slowly, deliberately, and comparatively little, the hyoidean elements would unite late in life, if at all. The complexity and modification of sounds depending largely upon the use of the tongue would also furnish sufficient reasons for early or late coössification.

§ 28. INDICES OF THE LONG BONES.

The indices of the long bones (Table LI) which have been taken are the antibrachial and the tibio-femoral. The measurements from which these were computed have been taken by means of the *planche ostéométrique* in use in France and according to the directions given by Topinard.[*] These directions require that all the bones except the tibia shall be so measured as to obtain their maximum length. The tibia is measured from the superior articular surface to the internal malleolus; thus the length of the intercondylar spine for the insertion of the cruciform ligaments is subtracted. The measurements have been taken with great care and are correct to a millimeter.

The indices are reckoned by means of the following formulæ: For the antibrachial index the length of the radius is multiplied by 100 and the product divided by the length of the humerus; for the tibio-femoral the length of the tibia is multiplied by 100 and the product is divided by the length of the femur.

Very few of the skeletons have complete sets of long bones. In many cases only one remains whole. Therefore, in order to obtain the greatest possible results, we have adopted the following plan:

Method I. We compute the indices from bones belonging to the same limb of the same skeleton.

[*] TOPINARD, op. cit., p. 1033.

Method II. Not having in a given case the material to do this, owing to the absence of one of the necessary bones, we use the calculated *average* length of the missing bone in place of the factor which the peculiar length of it would constitute if it were present. For instance, suppose we desire to calculate the antibrachial index for a limb of which the radius is missing, we multiply the average length of all the radii by 100 and divide the product by the length of the humerus; and if the humerus, instead of the radius, is missing, we multiply the length of the radius by 100 and divide the product by the average length of the humeri. Relatively corresponding formulæ are used for the posterior limb.

Thus we obtain two sets of figures, one which definitely states the relations of the bones in a given individual to whose skeleton both bones belonged, and one which states that a bone of a certain individual bears such and such a relation to the general average of certain related bones, whatever they may be, of his race.

In the synopsis (Table LII) giving the average osteometric indices the results obtained from the complete limbs by method I are given first, then those obtained by method II, namely, by the lengths of the bones compared to the averages. These two groups of figures, which sum up with very little difference, are then combined to give a general average for the race. In each of these groups of indices there are three subdivisions, one for the right side, one for the left, and one for the totals of both sides. The figures found at the bottoms of the columns of individual indices are the totals obtained from both methods. They reappear in the synopsis.

The extremes of the indices obtained by method II are preposterous and should be allowed no weight in discussing the variations in relative length of the segments of the limbs. The cause of the great variation in question is almost self-evident. They are from those cases where the individual was much above or much below the normal stature of the race. The cases where the index upon one side of a skeleton is calculated from two of its own proper bones, while upon the other side it is calculated from the relation of a bone's length to the average, often gives a startling difference between the right index and the left index, for which the above explanation accounts; but when we come to the average indices all these difficulties disappear and the figures obtained by method II come close to those obtained by method I. This we regard as sufficient justification for the adoption of method II as a means of increasing the number of individuals with whom our figures deal.

The reader is liable to think that he finds some obscurity with regard to the number of individuals concerned in the combined right and left or total index obtained by method II, and the same index obtained by methods I and II combined. Taking the antibrachial index as an instance, however (see *Synopsis of Indices*, Table LII), the cross line beginning with the words "computed by method II" and giving the number of total indices as 15 must not be read as if it ought to mean that there are 15 indices of each side combined to form a total of 15 indices of both sides but that the index derived from combining the aggregate of each side represents the average of a sum which consists of 30 factors.

A reference to the tables of antibrachial and tibio-femoral indices in Topinard's Anthropology[*] will show that the variation per cent of these indices is small. His minimum and maximum of the antibrachial index are 69.8 in a male Eskimo and 81.7 in a female Andamanese, respectively; hence only 11.9 per cent—this, be it noted, between individuals, not between racial averages. The tibio-femoral index varies from 78.6 in 9 male Esthonians to 89.0 in 1 female Negrito, or 10.4 per cent.

The maximum and minimum of series which contain five or more (individuals or limbs?)[†] are as follows: For the antibrachial index the maximum is 79.0 in 32 male African Negroes, the minimum is 72.4 in 26 female Europeans. For the tibio-femoral index the maximum is 84.4 in 10 African Negresses, the minimum is 80.2 in 5 male Chinese.

Continuing our study of Dr. Topinard's tables, we find that the sexual differences with regard to these indices are not great. As to the antibrachial index, the sexual difference ranges from 0.1 in Europeans to 3.0 in South Americans, the males having the higher index in each case. As to the tibio-femoral index, the sexual difference ranges from 0.3 in Europeans to 1.5 in negroes, the

---

[*] TOPINARD, *op. cit.*, pp. 1043–1045.
[†] Probably individuals. (See TOPINARD, *op. cit.*, pp. 1043–1045.)

males having the higher index in the first case and the females in the second. We believe, therefore, that the sexual difference is not sufficient to impair the value of the averages derived while combining the **sexes from a** relatively large series. Hence we do not state the sexes upon our tables. Indeed **it is less our policy** to investigate the sexual **and** other intraracial characteristics of this people than **to** accumulate facts and distinctions dealing with their place in the human series.

The relation between the antibrachial index and the tibio femoral index, as shown by Dr. Topinard's tables, may be, if we are permitted to borrow a term from craniology, *harmonic* or *disharmonic*. Thus both indices may be large or small; in that case the relation is harmonic, or one may be large and the other small; in this case the relation is disharmonic. Harmonic indices are the rule. Topinard calls attention to disharmonic indices in saying:

The Chinese, who have an elevated antibrachial index, have a low tibio-femoral index. The Bushmen, who **have** a low antibrachial index, have a relatively elevated tibio-femoral index.

We give (Table LIII) those of Dr. Topinard's figures, which **deal with 5 or more cases,** for comparative data. His own comment upon them that they "**rest upon too few cases**" should, however, be borne in mind.

It is considered by comparative anatomists that increasing length of the second segment of a limb as compared **with the first** segment is, when found in man, a low character. This opinion is grounded upon **the** knowledge of the relatively great length which the radii and tibiæ of anthropoid apes bear, respectively, to the humeri and femora. The criterion thus established places the Saladoans well **toward** the foot of the human scale. With regard to the antibrachial index they stand next **to the** bottom of the scale, between the Chinese, Annamites and Javanese above and the African negroes below, and removed three places, or 2.72 per cent, from the South Americans. With regard to the tibio-femoral index they stand at the bottom of the scale, next below the **South** Americans. These latter, therefore, we note in passing, seem to have quite disharmonic **indices of the long bones,** which the Saladoans certainly have not.

As will be seen from a glance at Table LIII, where we give extracts from Topinard's figures **and insert our own data** for the **Saladoans in** their proper places, the characters derived from the **study of the long bones must be** called discordant or "out of the series" by those anthropologists **who insist that all data of a** true scientific value shall group themselves in a scale having a European at the top, a **Chinaman in the middle, and a** Negrito at the bottom.

§ 20. THE SCAPULAR INDEX.

Owing to the greatly damaged condition of the skeletons, only fifteen adult scapulæ were in a sufficiently good state of preservation to be submitted to the measurements of length and width required for computing the scapular index. Of these, nine are right scapulæ and six are left scapulæ. The maximum index is 81.66; the minimum, 65.21, both found on the right side. The average for the right scapulæ is 71.42, and for the left 70.61. The general average is, then, 71.09 for the whole series.

Here again we find the Saladoans occupying a low position in the human series.

The following extract from Flower and Garson's [*] figures on the subject exhibit the position of **the Saladoans,** whose index we insert with reference to certain other peoples.

Races.	Indices.
6 Tasmanian scapulæ	60. 3
290 European scapulæ	65. 2
6 Bushmen scapulæ	66. 7
12 Australian scapulæ	68. 9
21 Andaman scapulæ	69. 8
15 Saladoan scapulæ	71. 0
6 Negro scapulæ	71. 7

But it is probable that the **distinctive numerical grades of** value of the scapular index differ from one another by so small degrees that large series must be measured in order to obtain figures

---

[*] FLOWER and GARSON: On the scapular index as a race character in man. Journal of Anatomy and Physiology, London, 1879.

sufficiently valuable to justify a conclusion. Indeed, if we rightly interpret the spirit of previous writers upon the subject, we should be inclined to believe that the series of Europeans given above is the only series large enough to be of undoubted value.

Broca* says:

* * * There is but a very slight difference between the human average and the averages of the great anthropoids, a difference so feeble that it disappears often when one considers, instead of the averages, the individual cases.

§ 30. TORSION OF THE HUMERUS.

Notwithstanding the opinion of Topinard, that the angle of torsion of the humerus gives "a good zoölogic character and a bad anthropologic character,"† we have determined it in all the humeri of this collection (41 in number), in which the necessary guiding marks as laid down by Broca‡ were found intact. Of this number 21 were from the right side and 20 from the left; but there were only 15 complete pairs. Of the latter 6 pertained to female skeletons, 5 to male skeletons, and 4 to skeletons of undetermined sex. Of 6 right unpaired humeri 2 were male and 4 of undetermined sex. Of 5 left unpaired humeri 1 was male and 4 of undetermined sex.

The degree of torsion was ascertained by a graphic system analogous to that employed by Lucae and Welcker,§ but by means of an apparatus different to theirs, which was devised by Dr. J. C. McConnell, of the Army Medical Museum, and is shown in Fig. 44.

It is a modification of the apparatus mentioned in § 3 and illustrated in Fig. 23. The periglyph (Fig. 24) is employed and the tracings are taken on varnished glass, inked and transferred to paper in the manner described in § 3. The frame (*a, a, a*) is much the same in both apparatus; but in the one now under consideration we have in the center of the frame a revolving stage with a clamp for holding the humerus.

Four long distinct parallel lines are drawn on the stage at right angles with its axis—one on each side of the clamp—on both the upper and lower surfaces, those on one surface being exactly vertical to those on the other.

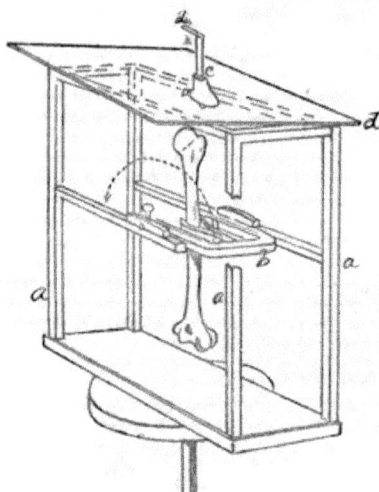

Fig. 44.—Apparatus for determining torsion of the humerus.

The angle of torsion is obtained in the following manner: Indicate on the humerus the axial lines of its two extremities. Secure the bone in a vertical position at the middle of the shaft by means of the clamp, let us say with the head of the humerus upwards at first; make a tracing of the head by means of the periglyph (*c*) on the varnished glass (*d*), being careful to include a tracing of the axial line and the parallel lines drawn on the stage. Invert the stage by turning it on its axle, inverting at the same time the bone and bringing the lower surface close to the varnished glass, where the outline with the axial and parallel lines may be sketched with the periglyph as before; ink the tracings and transfer to the paper in the manner described in § 3.

*P. Broca: Indices de Largeur de l'omoplate. Bulletins de la Société d'Anthropologie, Paris, 1878, p. 77.
† *Op. cit.*, p. 1048.
‡ La torsion de l'humerus. Revue d'Anthropologie, Paris, 1881. T. 2 serie, pp. 389 *et seq.*
§ Lucae: Die Stellung des Humeruskopfes zum Ellenbogengelenk beim Europäer and Neger, in Archiv für Anthropologie, 1866, i, p. 257 *et seq.*

In making the transfers to paper superimpose one sketch on the other in such a way that the axial lines shall cross or touch, and the parallel lines shall exactly coincide. Apply the protractor and read off the angle of torsion.[*]

In every case where we have applied this method we have, as a matter of record and identification, drawn the outlines of the extremities, an easy task; but it would have been sufficient to draw only the axial and parallel lines.

Figure 45 shows the character of the tracing. The outline of the head is broadened in order to distinguish it more plainly from that of the opposite extremity.

The general results of our measurements are shown in Tables LV to LVIII, inclusive, and in diagram shown below. Tables LIX and LX give the angles of other humeri in our Museum. A number of tables prepared by Broca and Manouvrier have been consulted which, though the measurements were taken by a different process, will, we believe, admit of comparison with our results. From all these sources the following facts are collated:

A statement of Broca's,[†] based upon abundant data, is that the average torsion is greater in females than in males, and his Table D shows that not only in the general average, but that in the average for each side the female exceeds the male, there being but one insignificant exception in the series of Californians. In this respect the Saladoans seem to be at variance with the rest of the human race. In Table LVIII it will be observed that humeral torsion in the males is greater on both sides, and therefore greater in the total average, than it is in the females.

Another conclusion of Broca's[‡] is that in nearly all the series (studied by him) the left humerus is, on an average, more twisted than the right; such, too, is the evidence of our general collection (Table LX) even with regard to American races. In the Salado skeletons, on the contrary, the average is almost the same on both sides, that of the right being slightly in excess of that of the left. Among the humeri in pairs, also, there is a slight excess on the right side. The variation, too, is greater on the right than on the left side in this series, the former showing both higher and lower angles than the latter.

In 75.8 per cent of Broca's series the maximum of torsion is on the left side.[§] Here again the Salado series ranges itself with the small minority. Not only the maximum but the highest

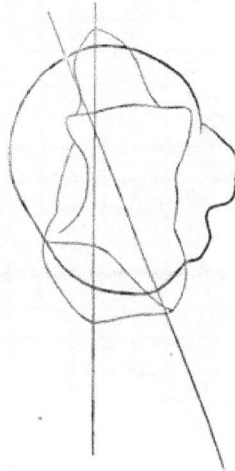

FIG. 45.—Tracing showing torsion of humerus.

three angles are found on the right side. It belongs to the majority, however, with regard to the minimum, which is on the right side as in 72.4 per cent of Broca's series.

In comparing the humeri of this series (Table LV) with those of our general series (Table LIX) we discover that three angles of the former (177°, 174°, 174°, all dexter) are higher than the maximum of any other race except the French, and that they are higher than several of the French angles. If we study this series in connection with Broca's Table C,[‖] in which is given a list of 29 series, comprising the most varied races in the world, the maximum angle of the Saladoans would still seem to have the same relative importance—standing next to the French.

The average torsion of the left humerus (159° 30′), the average of the right humerus (159° 45′), and the average of all the humeri (159° 30′ +) are higher than the corresponding averages in any series (representing more than one individual) of our collection except the French and Lapps.

---

[*] Journal of Anatomy and Physiology, vol. XXI, p. 536.

[†] La torsion de l'humérus. Revue d'Anthropologie, 2ᵉ Série, T. 10, Paris, 1881; pp. 577 et seq.

[‡] Loc. cit., p. 363.

[§] Loc. cit., p. 584.

[‖] Loc. cit., p. 583.

Diagram showing the difference between angles of torsion of right and left **humeri.**

RIGHT	1	2	3	4	5	6	7	8	9	10	11	12	13	14	15	16	17	18	19	20	21
LEFT		1	2	3	4	5	6	7	8	9	10	11	12	13	14	15	16	17	18	19	20
AVERAGE OF PAIRS			1	2	3	4	5	6	7	8	9	10	11	12	13	14	15				

Dotted line = right; single line = left; double line = both.

They are also higher than any of the averages in Broca's Table B,[*] except the modern Europeans and some of the ancient **Parisians**. They are higher than those of the French of the Polished Stone period. This Table B of Broca's shows 29 series, representing the most diverse races of the world, and is therefore an excellent basis for comparison.

§ 31. THE OLECRANON PERFORATION.

In the prevalence of the olecranon perforation the ancient inhabitants of the Salt River Valley **stand**, so far as we can learn, at the head of the human race. The following table shows the percentage of this anomaly in 24 series, of more than 15 humeri each, representing many different races and periods of time and arranged in order from the highest to the lowest percentage. It will be seen that the ancient Saladoans stand easily at the head of the list. We might **have** enlarged this table from our researches into the literature of the subject and by including smaller **series**, and yet have given no race precedence over the Saladoans.

TABLE 8.—*Showing percentages of olecranon perforation in different peoples.*

Number of humeri	Number of forations	Per cent.	Authority or collection.	Races or sources.
89	48	53.9	U. S. Army Medical Museum	Ancient Saladoans (Hemenway collection).
150	69	46	Bulletins de la Société d'Anthropologie. Paris, 1878, Vol. 1, p. 433.	Guanches, Canary Islands (Verneau).
39	.....	36.2	Topinard, Éléments d'Anthropologie Générale, p. 1016.	Yellow and American races.
32	.....	34.3	.....do.....	Polynesians.
80	.....	34.2	.....do.....	From Indian mounds in the United States (Wyman).
20	6	30	Private collection of Dr. D. S. Lamb	Dissecting-room specimens, mostly negro and mulatto.
62	17	27.4	U. S. Army Medical Museum	From Indian mounds in the United States.
122	.....	25.6	Topinard, Éléments d'Anthropologie Générale, p. 1016.	Guanches of Canary Islands.
156	.....	21.8	.....do.....	Dolmens and grottoes around Paris (Polished Stone period).
97	.....	21.7	.....do.....	African negroes.
61	12	19.6	U. S. Army Medical Museum	Ancient Cibolans (Hemenway collection).
28	.....	14.1	Topinard, Éléments d'Anthropologie Générale, p. 1016.	Melanesians.
39	.....	12.1	.....do.....	Dolmens of Lozère.
66	.....	10.6	.....do.....	Caverne de l'Homme-mort, Lozère (Polished Stone period).
388	.....	10.6	.....do.....	Dolmens of La Lozère (Polished Stone period).
288	22	7.9	U. S. Army Medical Museum	Pathological specimens, mostly from white soldiers.
27	2	7.4	Bulletins de la Société d'Anthropologie. Paris, Vol. v, p. 610.	From Chamont (Stone age).
16	1	6.2	U. S. Army Medical Museum	American negroes and mulattoes.
200	.....	5.5	Topinard, Éléments d'Anthropologie Générale, p. 1016.	Parisians from fourth to twelfth centuries.
96	5	5.2	U. S. Army Medical Museum	Modern American Indians.
150	.....	4.6	Topinard, Éléments d'Anthropologie Générale, p. 1016.	Parisians, Cemetery of the Innocents (Hamy and Sauvage).
218	.....	4.1	.....do.....	Parisians of the Middle Ages (Broca and Bataillard).
52	2	3.9	Revue d'Anthropologie. Vol. IX, p. 147	Europeans of America (Wyman, Peabody Museum reports).
30	0	0	Topinard, Éléments d'Anthropologie Générale, p. 1016.	Long barrows of England (Bronze age).

Perhaps some of the perforations were not counted. The bones of the Salado series, as before remarked, were very fragile, and the thin partition between the fossæ of the humerus was sometimes broken by accident. Pains were taken to distinguish between the natural and the accidental

---

[*] Op. cit., p. 582.

perforations. There was usually no great difficulty in doing this, as the margins of the former were smooth and bounded a fenestration, regular or subregular in shape—often oval, while the

irregular and fractured character of the margins of the latter was readily discernible. But it is probable that bones once perforated naturally were afterwards perforated post-mortem by fractures which included the natural fenestrations, or that the smooth edges of natural openings may have been abraded so as to give them the appearance of accidental openings; such cases would be excluded from the list.

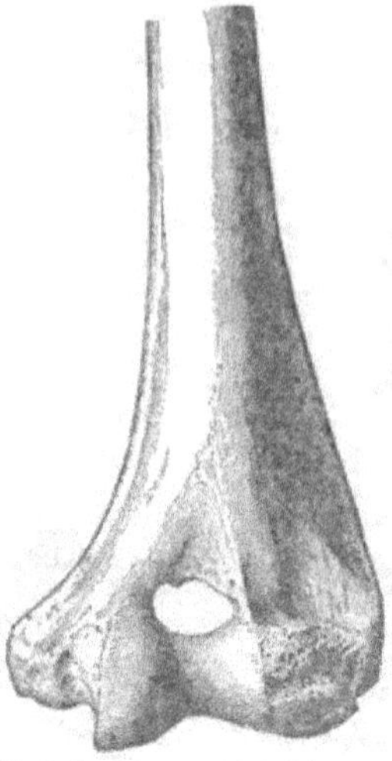

Not only is the perforation more common in this than in any other race, but, as far as our observations among the various series in the Army Medical Museum teach us, the number of large perforations is proportionally greater. Such, at least, was the impression gained during the examination; but we did not determine this by actual measurement. Fig. 46 represents, natural size, the lower extremity of a left humerus of an ancient Saladoan in the Hemenway collection. It exhibits an olecranon perforation 11 milimeters in length by 7 in width.

The following table of five series in the Army Medical Museum shows that the perforation is more commonly found on the left side than on the right; yet even in this particular the Saladoans differ much from the rest of the races. While with them, as with others, the perforation is more commonly found on the right side, the difference between the two sides is not so great. This is shown in the last column of the table.

The subject of the olecranon perforation has been so extensively discussed* that we deem it well to do little more than give the results of our studies of the Hemenway series

Fig. 46.—Lower end of humerus showing large olecranon perforation.

and other series in the Army Medical Museum, and indicate how our discoveries bear on the whole subject.

TABLE T.—*Showing percentages of olecranon perforation, on the right and on the left side, in different peoples.*

Race or collection.	Right.			Left.			Proportions of right to 100 left. Approximate.
	Number of humeri.	Number of foramina.	Per cent.	Number of humeri.	Number of foramina.	Per cent.	
Ancient Saladoans, Hemenway collection	43	19	44.1	46	29	63	70
Indian mounds, United States	35	7	20	27	10	37	54
Ancient Cibolans, Hemenway collection	30	2	6.7	31	10	32.2	20
Dr. Lamb's collection, mostly negro and mulatto	11	2	18.1	9	4	44.4	40
Pathological collection, mostly white soldiers	160	6	3.7	138	16	11.5	32
Total	279	36	12.9	242	69	24.3	53

We will attempt neither to cite the various theories which have been proposed to explain the nature and origin of the perforation, nor to quote the many arguments advanced to sustain those theories. We will merely announce that we are among those who believe that the perforation is not congenital but acquired; and that it has no connection with the rank a people may hold in the scale of races, but is the result of some mechanical cause connected with their occupations. We believe, furthermore, that it results from repeated and forcible extension of the forearm, in which the summit of the olecranon process of the ulna impinges against that thin bony partition which

---

* For a synopsis of the discussion and a bibliography of 43 titles, see "The Olecranon Perforation," by Dr. D. S. Lamb, in The American Anthropologist for April, 1890.

ordinarily separates the coronoid from the olecranon fossa of the humerus. The absorption of this partition and the consequent formation of a perforation connecting the two fossæ naturally follows.

Fig. 47 represents the anterior aspects of the distal extremities of both humeri from the skeleton of a young subject in the Salado series. The right humerus has a single large olecranon opening. In the left humerus the partition between the two fossæ is of a translucent thinness and is perforated by a number of small orifices which outline a space larger than the perforation in the right humerus. This left humerus is believed to present an olecranon perforation in the first stages of its formation. No other specimen of this character has been seen by us.

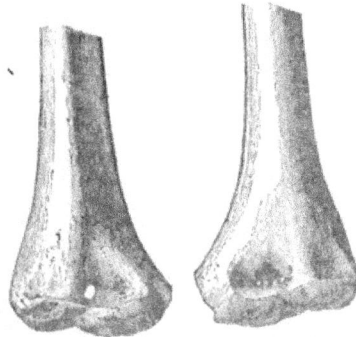

Fig. 47.—Lower ends of humeri showing olecranon perforations.

Our whole museum collection shows the perforation in two adolescents but in no infants. As far as we can learn the same fact has been observed with regard to children in other collections, and this is one of the facts on which rests the theory that the perforation is acquired and not inherited.

If it be granted that the perforation arises from mechanical causes and is the result of labor which requires repeated and forcible extension of the forearm, we need not search long to discover the existence of such labor among the aborigines of the southwest, both ancient and modern. The females of the modern pueblos are engaged during the greater part of their time in grinding corn, and they begin to perform this labor while they are yet very young. The grinding is done on a metate or large flat stone, by means of a smaller stone which is held in the hands of the operator and moved back and forth. The chief extension is made in moving the stone forward, and this requires the most forcible extension of the forearm. The motion is made chiefly by the muscles of the back. The discovery of numerous *metates* and upper grinding stones in the ruins of the Salado cities shows that the people practiced a method of grinding similar to that of the modern sedentary Indians of the same region. There were, no doubt, other labors which required great extension of the forearm, but this we believe was the most important.

Modern agricultural tribes of the North and East ground their corn in wooden mortars with wooden pestles; and in so doing made motions very different to those employed in operating with the metate.

Pruner Bey expresses the opinion that this peculiarity is, in the human race, to be found only in females, because all the humeri in which he noted the perforation were small. We can not say, for certain, that it is found only in female humeri, in the Salado series, because we can so rarely determine the sex of these skeletons; but it is not improbable that the perforation may be shown to occur more frequently among the females than among the males. Although the men did much hard labor of various kinds the work of grinding the corn was, in all probability, with the ancient Saladoans, as with the modern pueblo Indians, performed exclusively by the women.

That the perforation is not a peculiarity of females in all races is evidenced by the pathological series of the Army Medical Museum. In this series is a percentage of 7.9 perforations in 288 humeri, and these bones are, with few exceptions, derived from American soldiers of the Caucasian race. It is easy to conceive that many of our modern mechanical employments, such as that of the carpenter propelling the plane, in which the arm is forcibly extended, might cause the perforation we speak of. We have in our anatomical series the skeleton of a Frenchman showing the perforation on one side.

On the supposition that the perforation is produced by mechanical causes, we can account for its preponderance on the left side only by supposing that the left arm, in many occupations, is more frequently and forcibly extended than the right. For the majority of human manual tasks we are not prepared to demonstrate this, although we might do so in some instances. In the work of grinding on the metate, however, it appears that the left hand is used the more. When the grist is lifted from the trough and placed on the metate—and this is very frequently done—the right hand is employed while the left hand is not released from the grinding-stone.

§ 32. THE PELVIS.

Pelvic measurements have been practiced upon 19 articulated pelves besides one pair of innominate bones, 2 innominate bones of separate individuals with their corresponding sacra, 1 without sacrum, and 8 separate sacra (Tables LXI to LXVIII, inclusive).

The measurements are as accurate as could be hoped for in pelvimetry where landmarks are relatively quite indistinct.

No measurement has been permitted to originate with us. The series of 19 measurements are compiled from Garson[*] and Verneau.[†] Fritsch,[‡] Davis,[§] and Bacarisse,[‖] have also been consulted and the choice of each measurement determined by its frequency in use and its clear definition fully as much as by its apparent morphological utility. It was our orginal intention to extend the number of measurements to 21 by including a measurement of the height of the entire articulated pelvis and the subpubic angle; but although both these measurements have often been taken by investigators, we could not find sufficiently exact definitions to warrant our adoption of them.

The indices which have been calculated by different authors are very varied. In view of this fact, and also because all published series of measurements which we have examined deal with series which compared to **craniological series are** absurdly small, we have limited our indices to the two which Topinard especially **recommends,[¶] and** a few others which appear most useful in the discrimination of sex.

Verneau, however, seems **to base his discussion** of sex on anatomical differences and absolute measurements, while J. G. Garson and most other writers have given us practically no information concerning the male pelvis. Hence as we are dealing with an unknown people, indeed almost all American tribes are unknown to pelvimetricians, and a people of probably conspicuously small stature, we might very readily go astray in applying to any great extent the canons or results of European anthropometry.

With these considerations in view we have decided upon the following indices:

First.—The breadth-height index or relation of the maximum external width of the **pelvis at** the iliac crest to its maximum height, or, which is the same thing, the **maximum** length of the innominate bone.

$$\text{Formula:} \quad \frac{\text{Pelvic width} \times 100}{\text{Pelvic height.}}$$

Second.—Index of the superior strait.

$$\text{Formula:} \quad \frac{\text{Antero-posterior diameter of brim} \times 100}{\text{Transverse diameter of brim.}}$$

Third.—Index of the pubo-ischiatic depth.

$$\text{Formula:} \quad \frac{\text{Pubo-ischiatic depth} \times 100}{\text{Maximum width of superior strait.}}$$

Fourth.—Index of sacral length.

$$\text{Formula:} \quad \frac{\text{Sacral length} \times 100}{\text{Maximum width of superior strait.}}$$

[*] GARSON: Pelvimetry; Journal of Anatomy and Physiology, London, 1881–'82; pp. 106 et seq.
[†] VERNEAU: Le Bassin; Paris, 1875.
[‡] FRITSCH: Die Eingeborenen Süd-Afrika's, Breslau, 1872, Tabellé II.
[§] DAVIS: Thesaurus Craniorum, London, 1867, Appendix B.
[‖] BACARISSE: Du Sacrum, Paris, 1873; Thèse pour le doctorat.
[¶] TOPINARD: Éléments d' Anthropologie Générale, Paris, 1885, p. 1049.

Upon inspecting the pelves we find them forming two groups. In the one group are Nos. H. 6, H. 7, H. 14, H. 18, H. 19, H. 25, H. 41, and H. 72. These present all the ordinary characteristics of the male pelvis. In the other group we find Nos. H. 1, H. 5, H. 8, H. 10, H. 15, H. 36, H. 39, H. 45, H. 57, and H. 59. These represent females.

No. H. 96 is the pelvis of a very young person. The ilia, ischia, and pubes are not coössified. Hence, we do not attempt to determine its sex, and omit it from our calculations.

The data furnished by the breadth-height index and the index of the superior strait accord to the Saladoans a high place in the human series. With regard to the breadth-height index, both the males and the females stand at the top of the scale. With regard to the index of the superior strait, the females stand at the highest (arithmetically the lowest) end of the scale above the Europeans. The males occupy a medium position.

The other indices are of use in comparing the sexes, but we have never seen any comparative data concerning them in print.

We have prepared four ordinations, one for each index. These, especially the indices of pubo-ischiatic depth and that of sacral length, show very prettily the natural grouping of the sexes. (See Tables LXII to LXV, inclusive.)

NOTE.—In the table of measurements of the pelvis the abbreviation "5 v" is occasionally found after figures concerning the sacrum. This is used in cases where the sacrum consists of six vertebræ, to indicate that only five of them are measured.

LIST OF MEASUREMENTS.

1. *Conjugata externa.*—Antero-posterior maximum diameter of the pelvis: From the antero-superior part of the symphysis pubis to the summit of the spinous process of the first sacral vertebra.

2. *Crest width.*—Width between the crests of the ilia: Indicated by the greatest width between the external surfaces of the crests.

3. *Antero-superior spinal width.*—Width between the anterior-superior spines of the ilia: From the center of the most prominent part of one spine to the corresponding point on the other.

4. *Postero-superior spinal width.*—Width between the posterior-superior spines of the ilia: Measuring from the center of the most prominent part of one posterior-superior spine to a similar point on the other.

5. *Antero-posterior diameter of the brim.*—From the anterior-superior margin of the promontory of the sacrum to the most adjacent point of the symphysis pubis.

6. *Transverse diameter of the brim.*—Maximum width measured at right angles to the antero-posterior diameter.

7. *Antero-posterior diameter of outlet.*—Width between the center of the anterior-inferior margin of the body of the fifth sacral vertebra and the most adjacent point of the symphysis pubis.

8. *Transverse diameter of outlet.*—The maximum width of the pelvic outlet measured at right angles to the antero-posterior diameters of the outlet, between the most widely separated points, on lines passing parallel to the brim line from the spines of the ischia to the lower ends of the obturator foramina.

9. *Sciatic width.*—Minimum distance between the sciatic spines.

10. *Pelvic height.*—Maximum length of the innominate bone or pelvic height.

11. *Iliac breadth.*—Maximum breadth of the ilium.

12. *Height of iliac fossa.*—Height of the internal iliac fossa from the superior strait to the most elevated point of the iliac crest. (Upon the superior strait Verneau's *point de repère* is situated at the middle of the distance which separates the sacro-iliac articulation from the point which corresponds to the maximum transverse diameter of the strait.)

13. *Cord of the brim.*—From the sacro-iliac articulation to the symphysis pubis (at the level of the superior strait).

14. *Pubo-ischiatic depth.*—The distance between the upper surface of the pubis and the lower surface of the ischium, from the smooth level surface on the pubic side of the ilio pectineal suture above to the lowest part of the tuber ischii.

15. *Acetabulo-symphysial width.*—Width between the posterior margin of the acetabulum and the symphysis pubis.

16. *Sacral length.*—Vertical length of the 5 sacral vertebræ.

17. *Sacral breadth.*—Maximum breadth of first sacral vertebra.

18. *Width of sacrum at brim.*—Width of the superior strait at the reunion of the anterior face and the base.

19. *Inferior width of sacrum.*—Width of the sacrum below (at the inferior part of the auricular surface).

§ 33. THE COLUMNAR OR PILASTER FEMUR, *FÉMUR À COLONNE.*

We have studied this peculiar form of the femur, not by classifying the bones according to 6 different degrees as first recommended by Broca, but by finding an index as he later advises. (See Tables LXIX to LXXIII.) In order to obtain this index we took two transverse measurements of the diaphysis at its center—one antero-posterior, the other lateral; we multiplied the former by 100 and divided the product by the latter as directed by Topinard.[*] Our results, therefore, may be compared with a table given by Topinard. Our maximum index is 147.61. Our average indices are for 66 right femurs, 114.74, for 65 left femurs, 116.94, and for 131 femurs of both sides, 115.83 (Table LXXII). In 15 series which Topinard gives us, representing ancient and modern Europeans, Negroes, New Caledonians, and anthropoids, but three are higher than the Saladoan. These are: 1 nameless femur, 158; 1 femur from Cro-Magnon, 128, and 5 femurs from the Grand Canaries, 117.5. These series are all so small that they can not be compared with ours to good advantage. Indeed, Topinard has no series approaching ours numerically; his highest is 20 African negroes. We are not, then, able to judge with any degree of exactness where the Saladoans stand among the various human races and the lower orders of animals in respect to lateral compression of the shaft of the femur, and prominence of the *linea aspera*; but we may safely say that few, if any, races of men possess these peculiarities to a more exaggerated degree, and that few if any are further removed in these particulars from the anthropoids. Whatever, then, are the causes which produce the pilaster femur, they may be sought among the Saladoans.

It has been often observed among other races that the pilaster femur and the flattened tibia are associated features, and the Saladoans offer no exception to this rule. The flattening of the tibia is perhaps more remarkable among them than the lateral compression of the femur. We have some evidence, too, that, in this series at least, these features are associated in a **direct** though not symmetrical or constant ratio. This is shown in Table LXXIII, in preparing which we have selected 5 skeletons whose tibiæ showed the lowest indices, *i. e.*, the greatest lateral compression, and 5 other adult skeletons whose (normal) tibiæ exhibited the highest indices and the least lateral compression. For these 10 skeletons we have presented side by side the tibial and femoral indices, and computed averages for the two groups separately. It will be seen by consulting the table that the low tibial indices are accompanied by high femoral indices, and vice versa; in other words, the lateral compression of the femur is in a general way proportionate to the lateral compression of the corresponding tibia. Since the lateral diameter is employed as the dividend in computing the index of the tibia, and the **antero-posterior** diameter is so employed in the index of the femur, the indices of these bones bear an inverse relation to one another, *i. e.*, the narrower laterally the tibia, the lower the index; the narrower the femur, the higher the index. These observations lead us to the conclusion that whatever causes operate to produce the platycnemic tibia operate as well to produce the pilaster femur. Under the next section (§ 34) we consider these causes with regard to the tibia, because in that connection we fancy we can discover their operation more plainly.

§ 34. PLATYCNEMIA, OR FLATTENED TIBIA.

There is probably no single series of bones in any collection which offers better advantages for the study of platycnemia than the bones of the Salado. They belong to a race apparently very homogeneous, whose general habits of life are well understood, and they present this peculiar

---

[*] *Op. cit.*, p. 1019.

formation more constantly and in a higher degree than those of any other collection of which we have seen a record. Furthermore, the series is extensive.

In obtaining the index of the tibia for this study we have adopted the method of Broca; that is, we have measured the bone at the level of the nutrient foramen, have multiplied the transverse dimension by 100 and divided the product by the antero-posterior dimension. We have found in the Salado collection 116 tibiæ sufficiently preserved to admit of these measurements. Ninety of these, which were collected along with the skulls or other bones of the same skeletons, and which were conserved immediately on being disinterred, are given in Table LXXIV. Twenty-six of the tibiæ form a miscellaneous group; they were gathered singly and belong mostly to skeletons which in the earlier days of the work of excavation were allowed to disintegrate from exposure to the weather or were crushed under the feet of thoughtless visitors. Since many of this miscellaneous set are cracked and warped, we repose less confidence in their dimensions than we do in the dimensions of the series of 90; hence we devote to them a separate table (LXXV).

It is a recognized fact that the flattened tibia does not occur in childhood, but that the peculiarity is acquired as years advance. To include immature tibiæ in the general average may therefore be thought to improperly diminish the average of platycnemia and increase the average index. We have two skeletons in which there is an exostotic crest, apparently the result of unusual muscular traction, posteriorly near the junction of the perpendicular with the oblique line. This formation, on the other hand, by falsely increasing the antero-posterior diameter, may be thought to improperly increase the average of platycnemia and decrease the average index. In Table LXXIV we have noted under the head of "Remarks" all instances of these disturbing factors, and we have calculated averages both inclusive and exclusive of such instances.

If we take an index of 75 as representing a normal tibia (and this may be regarded as a low standard) we find but four adult tibæ in 116 which may be regarded as normal. The lowest American index we have seen recorded is one of 48 in a tibia from a mound in Michigan. This instance is mentioned by Jeffries Wyman,[*] who expresses the index by saying that the transverse diameter is 0.48 of the antero-posterior diameter. In this connection he never uses more than two decimal figures; consequently the index, if expressed in the manner adopted by us, might have been a fraction higher. We may safely say then that two tibiæ of the Salado series (those of H. 19) are as low as the lowest American known. Absolutely the lowest for all races we have seen recorded are 3 mentioned by Kuhff, one from *Caverne de l'Homme Mort*, having index 47, and two from the Grand Canaries, having indices 42 and 36.

The most extensive table of averages we have observed is one of 39 series given by Dr. Kuhff.[†] In this the lowest averages are in series having but 1 or 2 specimens in each, and they therefore do not admit of comparison with ours. The lowest average he gives in any series comprising more than 2 specimens is that of certain prehistoric bones ("Lehm de Kollwiller") from Alsace, of which there are 11 specimens. The average index is stated at 63; but Dr. Kuhff gives no decimals in his table. The average of all our 116 tibiæ is 63.54 (Table LXXV); the average of 90 of the more perfect part of the collection, as shown in Table LXXIV, is 62.71; while the average of 78 adults free from complicating exostoses is only 61.88. We may safely say that no series of equal size in any collection will be found to show a higher average grade of flattening of the tibia than our Salado series.

The most satisfactory explanation which we have noted of the origin of platycnemia is that of Manouvrier.[‡] We regret that we have never seen his original paper on this subject; we derive a knowledge of his work entirely from a review.[§] He has concluded from a careful study of the tibia in its anatomical relations that the flattening is entirely due to "lengthening and straightening of the postero-external surface of this bone; that is to say, of the surface of insertion of the posterior tibial muscle," and this lengthening and flattening, Mr. Manouvrier wisely main-

---

[*] Fourth annual report of the trustees of the Peabody Museum of American Archæology and Ethnology, Boston, 1871, pp. 21, 22.

[†] De la platycnémie dans les races humaines. Revue d'anthropologie, second series, vol. IV, p. 285.

[‡] Platycnemia in men and anthropoids. Memoirs of the Anthropological Society of Paris, 2d series, Vol. III.

[§] Revue d'anthropologie, third series, vol. IV, 1889, pp. 207–210.

tains, is due, not to the direct but to the inverse action of the muscle, produced under the influ-ence of repeated, almost constant work. He shows that the flattening is not similar to that observed in the anthropoids; that it results from the action of a different set of muscles; that it is not one of the "simian characteristics" which we are so prone to find in races whom we consider inferior to ourselves; that it is an evidence not of inferiority, but of superiority, since it is pro-duced under the influence of a cause essentially human.

This inverse action of the *tibialis posticus* is exerted when the foot is fixed and the tibia raised, as in the act of rising from a kneeling position. "This traction," says the reviewer, probably fol-lowing Manouvrier, "is produced in the upright position; more still in walking, above all up inclined planes, both in mounting and descending them, and infinitely more in running and jump-ing. It is, therefore, very probable that platycnemia should be found in great walkers, amongst the peoples of a varied country, living a savage life, hunting, etc. Children not presenting it shows it to be an acquired characteristic which is developed only at a certain age, under the influence of special conditions. We can explain this why it is less marked in the women, and why it presents in a given race very different individual degrees."

All the above suggestions as to causal activity are pertinent; but it seems to us that one of great importance remains to be made. When the *tibialis posticus* assumes the inverse action, the tibia becomes a lever of the second class, with the fulcrum at the ankle joint, the power at the insertion of the muscle, and the weight (which in ordinary cases is but the weight of the body and the clothing) at the knee-joint. There are three ways (besides frequency of impulse) in which the muscle that supplies the power may be called into increased action: First, by increasing the dis-tance through which the lever moves, as in climbing hills; second, by diminishing the time in which it moves, as in running and jumping; third, by increasing the weight, as in lifting and carrying heavy loads. Largely to the third way we are inclined to attribute the prevalence of platycnemia among various American races, including the Saladoans. The latter lived in a wide plain some distance (10 miles at Los Muertos) from the nearest mountains, which are neither remarkably high nor steep, and it is probable that, except for religious pilgrimages, they resorted rarely to these barren summits—as unproductive, no doubt, in ancient days as they are now. The Sala-doans were, then, not mountain-climbers. As they did not subsist to any great extent on game, their exercise in running was probably mostly confined to their sports. But they had no large domestic animals and were obliged to be their own burden-bearers. The burdens, too, were not dragged after them in vehicles, but were carried on the head or the back. Thus was the harvest brought home; thus were the materials collected and elevated to construct their tall houses, and the earth that was taken from their vast canals and reservoirs was carried out in baskets on the backs of men and women. The work done in this way by the Saladoans must have been enormous.

We have now in mind many facts connected with the customs of other peoples which tend to strengthen this theory, but we will not take the present occasion to mention them. To those who are considering the problem of platycnemia in Europe we would suggest that they inquire what effect the introduction of large beasts of burden may have had on the form of the human tibia, and what effect such occupations as those of the porter and the hod-carrier may have in preserving the flattened form to a limited extent to the present generation.

We have not seen elsewhere noted a feature that is apparent on a slight inspection of the bones of this series, namely, that there is a flattening of the fibula which corresponds with that of the tibia. This correspondence is general but not uniform, i. e., while no constant ratio can be shown to exist between the indices of associated leg bones, a very flat tibia is generally accom-panied by a very flat fibula, an average tibia by an average fibula, and a normal tibia by a normal fibula. The index of the fibula is usually less than that of its companion tibia. To elucidate these points we give a short table, in preparing which we have selected for illustration, from the Salado series, two very flat, two average, and two normal tibiæ. We have added one European tibia. All are from the right side.

TABLE LI.—*Flattening of tibia and corresponding fibula.*

Designation of skeleton.	Antero-posterior diameter of osseous of fibula.	Transverse dimension of fibula.	Index of fibula.	Index of associated tibia.
H. 19	19½	9½	40.87	48.75
H. 5	16½	10½	63.63	49.29
H. 15	13	7	53.84	62.26
H. 14	17	10	58.82	62.50
H. 96	14	9	64.28	75.43
H. 74	15	10½	70.00	79.63
Caucasian	15	10½	70.00	75.75

The measurements of the fibula were taken at the point where the maximum antero-posterior diameter was found. It is possible that, had some other point been selected for measurement, a ratio more nearly constant between the indices of the two bones might have been discovered.

The flattening of the fibula is accompanied by the following changes in the form of the bone: The entire shaft is twisted outward on its axis; the anterior portion of the internal surface is brought more to the front, almost forming a true anterior surface to the bone; the interosseous ridge, becoming more permanent and advancing to the front, divides the internal surface more sharply into two surfaces; the internal border becomes less distinct and allows the posterior surface, which largely loses its identity, to become merged with the posterior part of the internal surface. The bone is thus apparently compressed between the insertion of the *tibialis posticus* on the inside and the insertions of the *peroneus longus* and *peroneus brevis* on the outside. The two latter muscles are in their action adjuncts of the first. The "channeling" of the fibula noted by other observers is also found associated with these changes.

The columnar femur and platycnemia in various races.—Measurements have been taken to determine the relation of the various races as far as they are represented in our collections at the Army Medical Museum. This investigation has been fairly exhaustive and has embraced the large majority of all our accessible skeletons in good condition. In all 62 skeletons have been measured.

But even with all this number of individuals we find but two, or at most three, series which are sufficiently large to serve as the foundations of generalized assertions. These series are:

First, twenty-four Sioux Indians; second, twenty-three other Indians; third, six Negroes. (See Table LXXVIII.)

The reason why the Sioux are separated from the other Indians is simply because they form a sufficiently large series and not on account of any presupposed differences as to platycnemia and the pilaster femur. It might be as well in the present state of our knowledge to reckon all the Indians together and hence we have said above that the number of our comparative series of indices may be considered as either two or three.

The facts to be noted are that the Saladoans stand between our negroes and Indians with regard to the pilaster femur, while they possess tibia of a higher degree of platycnemia than any of the other races.

From the lists of the individual measurements and indices some curious data may be culled. We note the following as regards the columnar femur. (See Table LXXVI.)

Both the maximum and the minimum of the series are represented by bones of hunchbacks. The minimum index, that of the left femur of a white male, No. 5133, amounts to only 92.15. The maximum, which to the best of our knowledge is the highest index of the kind on record, is that of the right femur of the female negro, No. 5432, and amounts to 159.18. This surpasses by 1.18 per cent the femur of unknown origin which Topinard mentions as having the highest index of which he has ever been made aware.[*] There is nothing about this remarkable bone to suggest the action of disease. It is true that most of the arching forward, which we observe in all femora, is localized in this bone at about the junction of the upper and middle thirds. But this is a com-

---

[*] TOPINARD, *op. cit.*, 1019.

mon form. Indeed, judging from a plaster cast in the Army Medical Museum, we believe it to be the form of the celebrated femur of Cro-Magnon, whereof the index is 128. There is no suggestion of disease about the coxæ nor about the leg bones of this negress's skeleton, nor is there any peculiarity of the skeleton as a whole which in any conceivable mechanical means could have brought about such a result.

Two other hunchbacks, the one mentioned as having the minimum index of the section of the femur, and also No. 938, an Alaskan, show no such peculiarity.

In connection with platycnemia let us mention the Bannock male, No. 2133 (Table LXXVII). The indices of his tibiæ are 85.07 for the right and 93.75 for the left. These figures, while not the highest on record, are nevertheless very high, and show an entire lack of platycnemia if we may so express it. This is not what we should expect to find in an Indian skeleton, according to the facts learned in pursuing our investigation.

It happened thrice in our series of tibiæ that the nutrient foramen of a certain tibia was so very far out of normal place that it would falsify any measurement. Upon finding such a bone we would compare it to its fellow of the opposite leg of the same individual and measure it at a position corresponding to the level of the foramen in the latter. This is indicated on the margin of our tables.

HUMAN BONES

OF

THE HEMENWAY COLLECTION.

PART II.

THE SERIES OF CIBOLA.

PART II.—THE SERIES OF CIBOLA.

§ 35. THE SERIES OF CIBOLA.—ORIGIN, CONDITION, ETC.

The skeletons disinterred in the neighborhood of Zuñi are said to number about 200. **Thirty-five** of these, complete and incomplete, have been received at the Army Medical Museum. **The** others are stored in a house at Zuñi. These 35 were not selected for any scientific reason, but were packed and shipped because nearest at hand when the expedition was about to break up at Zuñi. They are mostly hard and in good condition, and present a striking contrast in this respect to the bones from the Salado Valley. They came mostly from the ruins of Heshota-uthla, which is about 13 miles in an easterly direction from Zuñi and further up the valley of the Zuñi River.

Heshota uthla was not one of the seven cities of Cibola. There is no doubt among those who have thoroughly investigated the matter that it was a shapeless ruin in 1540, when Coronado's army passed near its site. According to Zuñi tradition it was occupied in a remote antiquity by a people of their own race. Of this there is no evidence save tradition; yet for the present we place the remains of Heshota uthla along with the other remains from the same neighborhood in the **series** of Cibola. Archæological investigation shows that the people of Heshota-uthla had the **same customs**, arts, and general civilization as those of Cibola.

The ruin of Heshota-uthla, which was when inhabited a large, compact, many-storied pueblo, capable of sheltering a thousand or more people, lies close to the main wagon road from Zuñi to Wingate. Before excavations were begun it seemed to the untrained eye a natural heap of talus; the careful investigation of the scientific observer only revealed the fact that it was the ruin of a great edifice reared by human hands. Between the years 1880 and 1884 the writer frequently inspected this ruin alone and in company with Mr. Cushing and others, and it was then the general opinion that the heap of stones in sight represented the entire walls of the building from top to base. The recent excavations have shown that the loose stones were the débris of the upper stories only, the third and fourth perhaps, and that the first and second stories were buried from sight. The floors of the buildings were found at depths of 10 and 12 feet under the general surface of the ground—to such a depth had the surrounding soil accumulated by the washing of earth down from the neighboring hills and other natural causes since this pueblo was inhabited.

The skeletons in Heshota-uthla and in Hawien, as in the ruins of the Salado Valley, were found buried under the floors of the houses, but not with such care as in the latter place; no mud-walled graves were found, only ordinary holes in the earth, and the bodies were laid in all directions with relation to the points of the compass.

We have not given the same attention to the Cibolan collection as we have to the Saladoan. We have had less time to devote to it, and besides we have not thought it proper to give the Cibolan remains the fullest consideration until we should come into possession of the whole collection, which we hope to do at no distant day. We have taken some of the more important measurements and made sufficient study to enable us to draw a comparison between the skeletons of the Salado and those of the Zuñi Valley.

Some of the 35 skeletons come from Hawien, the Abacus of Coronado. This was about on the site of the present inhabited pueblo of Zuñi. The bones were exhumed near the main pueblo on the opposite bank of that narrow and inconstant streamlet known as the Zuñi River, and in the immediate vicinity of houses now occupied by certain extramural or outcast Zuñians.

§ 36. CEPHALIC INDEX. CIBOLA.

The antero-posterior shortening, which is such a marked feature of the Saladoan skulls (§ 6), is no less a marked feature of the skulls of Cibola. The tables (LXXXIII, LXXXIV) indicate even a greater shortening in the latter series. The shortest skull is broader than it is long, having an index of 100.69, a greater exaggeration of this shortening than is found among the Saladoans, whose highest index is 97.97. We have seen record of but few indices higher than the above of 100.69. The average cephalic index of this group (88.86) is higher than that of the Saladoans by a small fraction, notwithstanding that there are 5 Cibolan skulls longer than the longest Saladoan.

The minimum index, 74.54, which is dolichocephalic according to some authorities, belongs to a skull apparently normal and possibly of an alien race.

The supposed reasons for this shortening have already been declared.

§ 37. OCCIPITAL FLATTENING. CIBOLA.

This deformation, whatever be its significance, is the rule in the collection under consideration. Only 4 skulls, indeed, Nos. H. 201, H. 204, H. 221, and H. 229, can be called normal in outline.

Of the deformity of the remaining skulls it may be said that it can be most impressively explained by imagining it to have been made by a flat rigid surface moving in a plane vertical or tilted a little forward with reference to the antero-postero-horizontal plane of the skull, coming in contact with the occiput. Hence we find the flattening in the less notable cases involving only the most prominent part of the occiput, that is, from inion to lambda. Then we find a number flattened from inion to obelion, and lastly a few in which the whole occiput is affected. But this plane, while always approximately vertical to the aforesaid horizontal plane, may be either parallel to or at any angle with the transverso-vertical plane of the skull.

Hence the flattening may be strictly unilateral; or the flattening may affect both sides, but preponderate upon one; or the flattening may be bilaterally symmetrical. There are 10 skulls in which the flattening is nearly or quite bilaterally symmetrical. Eight skulls are flattened on the left side of the occiput, and twelve skulls are flattened on the right side of the occiput.

There is no skull exemplifying that occipital flattening wherein the occiput seems to have been in contact with a force pressing upward and forward. The resultant form is one in which the obelion is, or tends to be, the most posterior part of the skull, while the surface from the lambda to the inferior curved line, or even to the opisthion, forms a nearly continuous plane.

§ 38. VERTICO-LONGITUDINAL INDEX. CIBOLA.

The general remarks under the title "Vertical indices" (§ 11) made on the Saladoan skulls apply as well to the Cibolan, although we have placed on record for the latter only one vertical index, the vertico-longitudinal, whose factors are the greatest length and the basi-bregmatic height. (See Tables LXXXIII, LXXXIV.)

We found it possible to compute this index in 31 skulls only. The extremely short skull, H. 216, which gave such a high cephalic index, gave the still higher vertico-longitudinal index of 101.39, which was the maximum of the series; but it was not the normal skull with the lowest cephalic index (H. 209) that had the minimum vertico-longitudinal of 74.05. The variation of this index in the series of 31 is greater than in the Saladoan series of 39, and the average of the one series exceeds that of the other by 5 units.

The cephalic index and the vertico-longitudinal index of the Cibolan group are exactly the same in two cases* and they are within a unit of one another in 5 cases more.† The close correspondence of the maxima, the minima, and the averages in both indices may be seen by consulting Table LXXXIII.

§ 39. PLANE OF THE FORAMEN MAGNUM. CIBOLA.

In 27 skulls of this series we have been able to estimate the angle of Daubenton and the analogous basilar and occipital angles of Broca. (See Tables LXXXII, LXXXV.)

We found in the skulls of the Salado the highest expressions of these angles—higher than any previously on record, and we had thought that this might be a concomitant of the occipital distortion and due to pressure on the occiput in infancy, which caused the plane of the foramen magnum to incline more posteriorly. In the skulls of Cibola there is, to judge from the cephalic indices, as much of this flattening as among those of the Salado, yet the angles which indicate the inclination of the plane of the foramen magnum are not nearly so great in the former as in the

---

*Nos. H. 215 and H. 228.	†Nos. H. 202, H. 213, H. 216, H. 217, H. 226.

latter. They are the same among the Cibolans as they are in races with long heads who have no practices that result in flattening. For instance, according to Topinard,[*] the average angle of Daubenton is among Esquimaux, Hottentots, and Australians 6°, and among Javanese, Polynesians, and New Caledonians 7°, while among the Cibolans it is intermediate between these two, or 6½°. But it must be stated that Topinard omits fractions.

§ 40. PROCESSES AT BASE OF SKULL. THE INION. CIBOLA.

In general, the processes at the base of the skull are somewhat more prominent in the Cibolan than in the Saladoan skulls. This is particularly noticeable in the case of the inion, or, more properly speaking, the superior curved line in the region of the inion.

We have estimated the degree of projection of the inion indirectly from orthogonal drawings of the occiput. These drawings represent the most prominent points, whether they be in the sagittal plane or not. Hence if any part of the superior curved line of the occiput be more prominent than the inion proper, it is that which is represented upon the drawing and compared with the standard. This greater prominence of the superior curved line at one side of the insertion of the ligamentum nuchæ is met with several times (well marked in 8 cases; see Table LXXXVI) in the skulls under discussion. In general, the inion does not project much downward as a free process from the occiput, but is part of a large elevated bone area, quite distinct, and corresponding to the median part of the superior curved line. In short, it is the insertion of the trapezius muscle rather than the insertion of the ligamentum nuchæ which is exaggerated.

This greater prominence of the inial region of the Cibolan over the Saladoan skulls may be due to the fact that the pressure which flattened the skulls seems to have been exerted in the former entirely on a surface above the inion, while in the latter it was usually on a surface which included the inion. This remark must be taken in connection with what we have said in § 9.

In the Saladoan skulls none of the inia are more prominent than Broca's No. 1. In the Cibolan skulls (see Tables LXXXVI and LXXXVII) 12 out of 32, or three-eighths of all, correspond with higher numbers of Broca's scale. There is one which we consider as equaling his No. 4.

§ 41. THE PTERION. CIBOLA.

We have found in this series 40 pteria which admitted of measurement. They are equally distributed between the right and left sides. Eighteen skulls have both pteria intact. They are all of the form "pterion in H," but two are complicated with epipteric bones. (See Table LXXXVIII.)

The longest right pterion is 18ᵐᵐ; the longest left pterion, 24ᵐᵐ. There are two pteria of the right side measuring 9ᵐᵐ, but none measuring less. There are two of the left side measuring 7ᵐᵐ, and this is the minimum of the whole group. The average length of the right is 14.60ᵐᵐ, of the left 13.55 ᵐᵐ, of all 14.07ᵐᵐ. These averages are higher than those of Salado.

In the Cibola, as in the Salado group, there are but two pteria less than 8ᵐᵐ in length, but, as the Cibolan series is greater, it shows a smaller percentage, which is only 5.

Placing the above figure along with Anoutchine's tables, previously quoted, we find that the Peruvians have of all races the smallest percentage (3.4) of pteria—less than 8ᵐᵐ in length; that the Cibolans come next, with 5 per cent; the Saladoans third, with 6.5 per cent; the "People of the Caucasus, Turkestan, and Turko-Finnish" fourth, with 6.9 per cent, and that all other races have higher percentages, the highest being the Australians and Tasmanians, 24.6 per cent.

Like the Saladoans, the Cibolans show no frontal apophyses at the pterion, and they show but two epipteric bones (5 per cent), less even than the Peruvians, whose per cent (6) is the lowest on Anoutchine's table of 10 series.

In making this comparison it should be remembered that both of the series described in this report are much smaller than any of Anoutchine's, his lowest Australians and Tasmanians being 102.

---

* TOPINARD: Op. cit., p. 814.

Anoutchine gives in another table a list of 12 series of diverse races, in which there are pteria less than 3ᵐᵐ in percentages, ranging from 0.5 to 8.2 (Chinese). Since the Peruvians do not appear on this table, we presume their percentage is zero. Such is the case with the Cibolans, but, as before stated, we found one of these small pteria in a Saladoan skull.

§ 42. INCA BONE AND KINDRED FORMATIONS. CIBOLA.

As we have not secured illustrations to show for this series the anomalies involving the superior angles of the vertical portions of the occipital bone, we have prepared a list of these anomalies, 12 in all, with a detailed description of each case.

List.—H. 203: In the left limb of the lambdoid suture there are Wormian bones, one of which sends a process across the apex.

H. 206: A typical *os apicis* 32ᵐᵐ high by 51ᵐᵐ broad. It has Wormian bones at its lower angles.

H. 207: A very curious multiple apicial bone reaching to within 3ᵐᵐ of the obelion. It consists of nine principal and many smaller portions. The whole group is 43ᵐᵐ high and 51ᵐᵐ wide, and might by some be considered an *os Inca*. The bone forming the apex is 21ᵐᵐ by 17ᵐᵐ.

H. 210: Fine tortuous Wormian bones in both limbs of the lambdoid suture. There is one of this set at the apex, a small irregular ossicle, which might be regarded as an *os sagittale*.

H. 212: A large compound bone at the apex, mostly to the right of the median line. The lower part of it, about 40ᵐᵐ by 32ᵐᵐ, is partly coössified to the rest of the occipital. The upper part, about 22ᵐᵐ by 9ᵐᵐ, forms a small apicial bone. A few Wormian bones of small size complicate the lambdoid.

H. 213: A row of medium-sized Wormian bones in each limb of the lambdoid suture; one of these bones is on each side of the apex; 12ᵐᵐ above the apex is an *os sagittale* 18ᵐᵐ by 9ᵐᵐ.

H. 218: An apical bone in two parts slightly coössified. The total size is 25ᵐᵐ high and 47ᵐᵐ wide.

H. 223: A row of medium-sized Wormian bones, all of remarkably simple outline, in each limb of the lambdoid suture. One of these bones situated at the apex measures 14ᵐᵐ by 16ᵐᵐ.

H. 227: A large typical Inca bone 46ᵐᵐ high and 73ᵐᵐ wide. Above it is a small bone 7ᵐᵐ by 10ᵐᵐ, which may be regarded as an *os sagittale*.

H. 231: A triangular *os apicis*, 27ᵐᵐ by 53ᵐᵐ.

H. 232: A row of very irregular Wormian bones occupies the lambdoid suture throughout from the left asterion almost to the right; one of these bones, situated in the median line, is about 8ᵐᵐ by 14ᵐᵐ, and may be considered an apicial bone.

H. 233: A row of Wormian bones occupies the upper half of the right limb of the lambdoid suture; one of these, 13ᵐᵐ by 14ᵐᵐ, touches the median line at the apex.

From the above list we learn that there are in this series the following anomalies: 1 typical Inca bone, H. 227; 5 typical apicial bones, Nos. H. 206, H. 207, H. 212, H. 218, H. 231; 6 doubtful apicial bones, Nos. H. 203, H. 210, H. 213, H. 223, H. 232, H. 233, or 11 apicial bones of both classes—12 anomalies in all. The above numbers give us, in a series of 35, the following percentages: Inca bone, 2.85 per cent; true apicial bones, 14.28 per cent; doubtful apicial bones, 17.14 per cent; both classes of apicial bones, 31.42 per cent; total of all anomalies 34.28 per cent. From these percentages, from those given in paragraph 18, and from percentages obtained from our own collection we have prepared the following table:

TABLE W.—*Frequency of Inca bone in various peoples.*

Races.	Complete *os Inca*.	True apicial bones.	Doubtful apicial bones.	Apicial bones of both classes.	All anomalies.
Saladoans	5.68			18.1	26.04
Peruvians	5.46	10.5?		10.5?	17.63
Cibolans	2.85	14.28	17.1?	31.42	34.28
Americans not Peruvians	1.30	5.63?		5.63?	9.75

While this series is too small to enable us to institute a perfectly satisfactory comparison between it and others, we have nevertheless obtained data sufficient to allow us to conclude that, with regard to this class of anomalies, the Cibolans are in close relation to the Saladoans and the Peruvians, and widely separated from other American races and from the rest of the world.

§ 43. NASAL CHARACTERS. NASAL INDEX. CIBOLA.

The average nasal index is nearly the same in both the Saladoan and the Cibolan series; that of the former being 51.66, and that of the latter **51.88** (Table LXXXIV). The remarks, therefore, which apply to the one apply as well to the other. In respect to the maximum and minimum of this index, also, the two series correspond closely. The maxima are: For the Saladoan 61.11; for the Cibolan 60.46. The minima are: For the former 44.23; for the latter 45.09.

The character of the lower border of the nasal aperture or *échancrure* of the Cibolans seems from such evidence as we possess to be inferior only to that of the Europeans and Saladoans. For the two highest classes A + A′ (see Table LXXXIX) their percentage is 38.23. The lowest class, E, simian gutter, has no representative. Over one-third of the series belong to class B. The relation with regard to this characteristic, which the Cibolans sustain to other races, will be seen by comparing Table LXXXIX with the tables in § 24.

§ 44. TORSION OF THE HUMERUS. CIBOLA.

The average angle of torsion (154.27°) of all the humeri, 48 in number, of this series is higher than that of any race recorded by Broca except the Mexicans (155°) and the Europeans. As his Mexican series numbers only 2 it is scarcely worthy of being cited in comparison. Excluding the Mexicans, the Cibolans follow in respect to this feature next after the Saladoans and Europeans in the category of the human race, as far as we have seen the record. They are widely separated from other American races. (See Tables XC and XCI.)

Like the majority of mankind, and unlike the Saladoans, the Cibolans have the maximum angle of torsion on the left side. Not 1 but 5 angles on the left are higher than the highest angle on the right. The minimum is on the right side and there are 3 angles of the right lower than the lowest of the left.

The maximum angle of torsion of the Cibolans (178°) is higher than the maximum angle of Saladoans and, as far as Broca's tables inform us higher than the maximum of any people except the French. But the average of the highest 3 angles (173°) is not so great as the average of the highest 3 Saladoan angles (175°).

The average of all the left humeri, 23 in number, is 159.20°, while that of all the right humeri, 25 in number, is but 149.40°, a difference in favor of the left of nearly 10°. This is a higher difference than exists in any one of Broca's series, which represents more than two bones, except the Arabians and Kabyles + El Goleah, in which the difference is 10.27°.

The variation is greater on the left than on the right; on the one side it has a range of 35°, on the other a range of 20°.

§ 45. THE OLECRANON PERFORATION. CIBOLA.

The Cibolans present this anomaly in a much less degree than the ancient people of the Salt River Valley, the so-called Mound-Builders, the Guanches, and other peoples. The perforations appear in only 19.6 per cent of the humeri of the Cibolans, while the humeri of the Saladoans show 53.9 per cent.

The ancient people of the Zuñi Valley, no doubt, ground their corn in the same manner as did the ancient inhabitants of the Salt River Valley, and it may very pertinently be asked why the humeri of the former are not so often perforated as those of the latter. Retaining the hypothesis before mentioned that the method of grinding corn was an important factor in producing the olecranon perforation, we account for this difference by supposing that the Cibolans subsisted less on corn, and hence had less occasion to grind it than their more Western congeners. The land around Zuñi is not nearly so prolific as that of the Salt River Valley, the climate is colder, and agriculture is far less remunerative. The mountains adjacent to Zuñi, heavily timbered, abound in game, and it is probable that the ancient Cibolans lived more by the chase and less by agriculture than the ancient Saladoans.

TABLE I.—*General Measurements.*[*]—*Salado.*

1. Special number.

2. Museum number.

3. Age, in the 6 periods of Broca: 1st, 0 to 6 yrs; 2nd, 6 to 14 yrs; 3rd, 14 to 25 yrs; 4th, 25 to 40 yrs; 5th, 40 to 60 yrs; 6th, 60 yrs up.

4. Sex: M for male, F for female; ? for doubtful.

5. Capacity in cubic centimeters.

6. Horizontal length: from glabella parallel with horizontal plane to a perpendicular tangent to maximum occipital point. Frankfurt 1.

7. Greatest length: from the glabella to the maximum occipital point. Frankfurt 2. Topinard 1. p. 754.

8. Metopic length: from the metopion to the maximum occipital point. Frankfurt 3. Topinard A.

9. Greatest width: perpendicular to sagittal plane (not over mastoid process or at posterior temporal ridge). Frankfurt 4. Topinard 2.

10. Biasterie or maximum occipital width. Topinard B.

11. Bijugular or inferior occipital width. Topinard C.

12. Bimastoid width or width of the cranial bases: distance between the ends of the mastoid processes. **Frankfurt 13 a.**

13. Inferior subtemporal width: from one subtemporal point to the other. Topinard E.

14. Two frontal widths: 1st, smallest frontal, Frankfurt 5. Topinard 4; 2nd, maximum frontal or greatest of frontal bone, Emil Schmidt.[?]

15. Two auricular heights: 1st auricular height, Frankfurt 8. 2nd auxiliary auricular height, Frankfurt 9.

16. Horizontal circumference: above the superciliary ridge and over the most prominent part of the occiput. **Frankfurt 14; Topinard 5.**

17. Two Divisions of the horizontal circumference: 1st, anterior; 2nd, posterior. Separated by supra-auricular curve. **Topinard G.**

18. Sagittal circumference: from nasion to opisthion. Frankfurt 15.

19. Three divisions of the sagittal circumference: 1st, frontal; 2nd, parietal; 3d, occipital. Topinard F.

20. **Two** vertical circumferences: 1st, vertical circumference perpendicular to horizontal plane. Frankfurt 16. 2nd, supra-auricular curve.

21. Two dimensions of the foramen magnum: 1st, length in sagittal plane; 2nd, width perpendicular to sagittal plane.

22. Zygomatic width: greatest distance between the zygomatic arches. Frankfurt 18. Topinard 8.

23. Bimalar width: from external extremity of small fronto-malar suture to same point opposite. Topinard 9.

24. Facial width: from inferior extremity of maxillo-malar suture to corresponding opposite point. Frankfurt 17. Topinard 11.

25. Inter-orbital width: Distance from one dacryon to the other. Topinard H.

26. Two facial heights: 1st, total, nasion to lower border of inferior maxilla; 2nd, upper, nasion to alveolar point. Frankfurt 19, 20.

27. Two nasal dimensions: 1st, length, nasion to upper border of nasal spine; 2nd, maximum width. Frankfurt 21, 22. Topinard 17, 18.

28. Two orbital dimensions: 1st, dacryon to opposite margin in grand axis; 2nd, greatest height perpendicular to preceding. Topinard 19, 20.

29. Two palatal dimensions: 1st, length of bony palate; 2nd, median width of palate. Frankfurt 27, 28; Topinard O, L (less exact).

30. Two palatine widths: 1st, posterior. Frankfurt 29; Topinard M; 2nd, anterior, between canine and second incisor. Topinard K.

31. Depth of palatine arch: maximum, from alveolar edge, avoiding posterior palatine foramen. Topinard.

32. Two alveolar widths: 1st, external maximum, taken at level of malar region; 2nd, external posterior. Topinard.

33. Superior facial projection, or projection of the ophryon with regard to the alveolo-condylean plane. Topinard I.

34. Two widths of the lower jaw: 1st, external bicondylar; 2nd, external bigonial. Topinard 12.

35. Two dimensions of the ramus: 1st, height from angle to upper edge of condyle; 2nd, width, perpendicular to height. Topinard Q.

36. Basilo-mental radius: basion to mental point. Topinard S.

37. Superior alveolar radius: basion to alveolar point. Kollmann's "length of profile of face." Frankfurt 26; Topinard.

38. Nasal radius: basion to nasion. Topinard V, ?. "Length of cranial basis." Frankfurt 16.

39. Intersuperciliary radius: basion to glabella. Topinard W.

40. Metopic radius: basion to metopion. Topinard X.

41. Two vertical radii: 1st, "basilo-bregmatic diameter," **Topinard 3**, "auxiliary height," Frankfurt 7; 2nd, "Entire height," after Virchow.

42. Obelic radius: basion to obelion. Topinard Y.

43. Inial radius: basion to inion. Topinard Z.

44. Occipito-alveolar length: from maximum occipital point to alveolar point. Topinard 22.

45. Occipito-spinal length: from maximum occipital point to inferior border of nasal aperture. Topinard 23.

46. Two cranial projections: 1st, anterior or prebasilar; 2nd, posterior or post-basilar; both alveolo-condylean plane. **Topinard d, e.**

47. Profile angle (German): angle of naso-alveolar line on audito-orbital plane. Frankfurt.

48. Angle of Daubenton: sub-orbito-opisthiac line with plane of foramen magnum.

49. Two other angles of plane of foramen magnum: 1st, occipital angle with naso-opisthiac line; 2nd, basilar angle with naso-basilar line.

50. Cephalic index: No. 9 × 100 ÷ No. 7.

51. Vertico-longitudinal index: No. 41, 1st, × 100 ÷ No. 7.

52. Index of the foramen magnum: No. 21, 2nd, × 100 ÷ 21, 1st.

53. Facial index of Virchow: No. 26, 1st, × 100 ÷ No. 24.

54. Upper facial index of Virchow: No. 26, 2nd, × 100 ÷ No. 24.

55. Facial index of Kollmann: No. 26, 1st, × 100 ÷ No. 22.

56. Upper facial index of Kollmann: No. 26, 2nd, × 100 ÷ No. 22.

57. Nasal index: No. 27, 2nd, × 100 ÷ No. 27, 1st.

58. Orbital index: No. 28, 2nd, × 100 ÷ No. 28, 1st.

59. Palatine index: No. 29, 2nd, × 100 ÷ No. 29, 1st.

60. Gnathic index: No. 37 × 100 ÷ No. 38.

---

[*] All measurements in this table, not otherwise specified, are given in millimeters. [?] For further particulars as to measurements see § 2.

TABLE 1—Continued.

1	H, 1.	H, 2.	H, 3.	H, 4.	H, 5.	H, 6.	H, 7.
2							
3	4th	4th	6th	3d	4th	5th	4th
4	F.	f	f	f	F.	M.	M.
5							1330cc
6	154	157	157	147	168	159	162
7	154	169	157	150	168	156	162
8	150		152	139	160	158	152
9		144	151	145	147	149	139
10		106		98		103	100
11	77	86	85	81	87c	80c	72
12	100	103	110	100		110	100
13	83	86			88		85
14	84	86; 119	97	82; 108	100; 123c	84; 119	93; 121
15	115; 115	125; 125	116; 116	111; 111	120; 120	120; 120	117; 117
16		485	491	459	504	486	480
17		298; 275		205; 254	230; 271	246; 240	217; 263
18	330	319	339	329	354	342c	332
19		118; 126; 165		119; 112; 98	123; 113; 116		120; 111; 101
20		345; 319	322	325; 295	340; 320	345	334; 311
21			27;	—; 29	36; 35c		35; 30
22	126		139				131
23	97	106	105	95	110	90	106
24	93	95	97	98	113	98	100
25		25	234	184		90	22
26	109; 68	—; 65	—; 71	110; 66	114c; 68	—; 68	—; 66
27	47½; 23	47; 24	31; 27	46; 25	50; 26	30; 23	30; 24
28	41; 34		30; 34	37; 36	41; 33	30; 32	40; 28½
29	56; 37	50; 36	50; —	43; 40	57; 38		52; 38
30	37; 25	37; —	—; 24	40; 26	30; 24		41; —
31	13	13		14	18		13
32	62; 49	62; 30	—; 52	67; 51	65; 49		62; 51
33	28	28	12	28	19c	15	19
34	114; 97		—; 97	—; 89	122; 95	—; 101	116; 99
35	56; 35	68; 32	55; 30	57; 29	90; 32	64; 32	56; 29
36	110			111	108		
37	100	99	92	98	99	105	97
38	99	97	101				
39	107	105	109	102	112	118	111
40	115	121	121	109	119	127	119
41	133; 134	144; 148	—; 138	128; 153	140c; 141c	—; 142	133; 136
42		141		122	135		123
43	54	76	65	57	68	85	64
44	186	199	180	201	203	176	193
45	174	187	168	184	180	170	183
46	105; 78	102; 88	91; 81	100; 86	101; 91	106; 72	99; 87
47	79½c	78c	86½	77c	87c	86c	84c
48			15		11	10	14c
49			24; 31½c		29; 26c	19; 24c	27½; 36½c
50		90	96, 17	96, 90	97, 50	93, 71	85, 90
51	86, 36	90	88, 59	85, 33	83, 33		85, 35
52					97, 22		85, 71
53	111, 65			112, 24	110, 67		
54	72, 72	68, 42	73, 71	67, 34	66, 91	69, 32	66, 00
55	89, 30						
56	53, 96		51, 43				50, 38
57	48, 12	51, 06	52, 94	51, 34	52, 00	46, 00	48, 00
58	82, 92		87, 17	97, 29	85, 36	84, 61	88, 75
59	65, 48	73, 60		81, 63	66, 66		73, 07
60	101, 01	102, 06	92, 07	110, 11	95, 11	95, 45	97, 00

* This is at the maximum occipital point; the circumference at lambda is 463.

TABLE I—Continued.

1	H, 8.	H, 9.	H, 10.	H, 11.	H, 12.	H, 13.	H, 14.
2							
3	5th	6th	4th	4th	5th	4th	4th
4	F.	?	F.	?	?	?	M.
5			1310cc			1179cc	1510cc
6	161	152	155	157	176	152	166
7	163	151	158	157	176	159	106
8	157	143	146	145	168	142	162
9	144c	132	143	138	139	134	147
10	104	106	103	111	149c	100c	99
11	76	85	87	83c	86c		82
12	109	112	115		108	107	110
13	80	86	82	83	86	84	86
14	91; —	89; 108c	88; 124	86; 111	91; —	84; 111	83; 118
15	119; 119	120; 120	123; 123	114; 114	116; 116	115; 115	122; 122
16	487	451	472*	463	404	455	488
17	213; 274	204; 267	204; 272	231; 230	229; 265	207; 248	229; 280
18	350	349	342	326c	350c	335	353
19	120; 138; 102	116; —; —	110; 133; 99	112; 108; 105	132; —; —	111; 112; 112	128; 123; 102
20	334	301; 283	336; 314	328; 294	332; 313	316; 292	350; 319
21	32; 30	34½; 33	35; 31			—; 29	36½; 30
22	131	126	126	129		127	128
23	105	103	104	101	109	101	101
24	90	101	93	106c	101	100	98
25	28		22	21	26	22	194
26	—; 66	—; 69	—; 71	118; 71	—; 66c	—; 67	125; 78
27	47; 24	49; 22½	49; 23½	49; 24	—; 23½		53; 24
28	38½; 34½		42; 37	39; 37	40; 38	39½; 35½	37; 30½
29	51; 38	51; —	51; 36	51; 36	52; —		57; 41
30	38; —	36; 23	38; 22	39; 25			40; 25
31	12		18	14			18
32	52; 45	—; 49	62; 49	60; 52	—; 46	—; 51	62; 47
33		22	18	27	12	17c	25
34	115; 91		114; 95				116; 94
35	59; 33	56; 31	59; 27	61; 30½	63; 31	51; 30	63; 29
36				100			105
37	94	96	96	98		94	99
38				98		100	101
39	106	109	110	109		109	114
40	113	116	117	115		116	121
41	—; 140	135; 136	145; 147	133; 136		140; 145	138; 142
42		132				135	134
43	66	71	69	64		75	71
44	193	184	203	199	197	191	106
45	185	168c	186	185	193		184
46	95; 80	97; 82	96; 92	100; 91		39; 70	100; 98
47	84½°	78°	80°	80°		81°	86°
48	101°		18½°				154½°
49	211°; 281½°		30°; 33°	30°; 33°			20°; 53½°
50	88.07	87.41	30.50	87.89	78.97	87.58	88.33
51		89.46	91.77	84.71		91.50	83.13
52	93.75	95.65	88.57				82.19
53				118.00			127.55
54	72.92	68.31	76.94	71.00	38.54c	67.90	79.59
55				91.47			97.65
56	50.58	54.76	56.34	55.03		52.75	60.93
57	51.06	45.91	47.95	48.97			45.28
58	89.61		88.09	94.87	95.00	89.87	98.64
59	74.50		70.58	70.58			71.02
60	95.91	91.83	95.00	100.00		98.00	98.01

* The maximum circumference is 476.

TABLE I—Continued.

1	H, 15.	H, 16.	H, 17.	H, 18.	H, 19.	H, 20.	H, 21.
2	....	....	....	....	....	....	....
3	4th	4th	3d	5th ?	5th	4th	5th
4	F.	?	?	M.	M.	?	F.
5	1150cc	....	....	1290cc	1330cc	....	....
6	158	158	160	163	175*	180*	160
7	159	150	164	169	171	179	159
8	131	151	150	160	167	177	147
9	134	134	145	145	148	....	147
10	97	100	98	100	108	101	112
11	73	78	74	74	86	....	....
12	98	101	94	110	115	115c	117
13	85	81	87	87	94	85c	....
14	86; 113?	78; 109	98; 124	92; 112	100; 125	92; ——	96; ——
15	110; 110	....	117; 117	120; 120	126; 126	127; 127	114; 114
16	467	472	489	496	504	504	476
17	222; 245	208; 264	228; 261	221; 275	238; 266	244; 263	236; 250
18	338	354	341	353	354	376	323
19	120; 107; 111	115; 136; 106	121; 113; 107	130; 119; 104	128; 121; 105	——; ——; 111	113; 109; 101
20	315; 290	——; 306e	343; 320	335; 316	340; 320	333; 312	315; 300
21	32; 26½	31; 29½	32; 28	34; 29	36; 34½	....	....
22	126	121	127	137	146	139	139
23	100	96	*102	104	117	106	109
24	91	93	100	98	99	96c	102
25	27½c	20½	25	25	24	25	28½
26	118; 68	99c; 61	100; 66	118c; 69	120; 72	126; 71	——; 68
27	51; 23½	45½; 25	49; 26	50; 27	52; 28	45; 27	53; 25½
28	37; 30	37; 34	38; 34½	37; 33	42; 36	38½; 30½	39; 23½
29	62; 33	49; 40	54; 40	50; ——	52; 44	55; 36	54; ——
30	34; 24	39; 24	40; 24	——; 25	44; 27	36; ——	....
31	15	14	18	....	19	17	....
32	——; 47½	62; 47½	63; 48	——; 51	70; 54	66; ——	——; 47
33	24c	13	13	13	13	18	....
34	....	115; 91	113; 82	117; 96	131; 106	....	121; 103
35	60; 31½	61; 28	57; 30	66; 31	68; 33	65; 36	65; 31½?
36	102	101c	99	104c	110	....	....
37	88	88	89	95	98	....	....
38	91	94	98	107	109	....	....
39	101	104	107	120	121	....	....
40	110	112c	113	130	128	....	....
41	127; 129	....	136; 137	146; 145	146; 147	....	....
42	129	....	....	130	129	....	....
43	68	68	63	72	68	....	....
44	186	192	193	193	194	193	184
45	174	183	185	182	185	183	177
46	91; 93	89; 89	89; 93	98; 87	102; 91	....	....
47	84°	81°	88½°	88°	89°	82½°	86°
48	4½°	7½°	17°	20°	14°	12½°	....
49	13½°; 18°	19°; 25°	29°; 38½°	30½°; 40°	25½°; 32½°	23°; 30½°	....
50	84.23	84.27	88.41	85.79	88.54	76.53	92.45
51	79.87	....	82.92	86.39	85.38	....	....
52	82.81	95.16	87.50	85.29	95.83	....	....
53	120.67	106.43	109.09	120.40	121.21	131.25	....
54	74.72	65.59	64.00	70.40	72.72	77.08	66.66
55	93.65	81.81	85.86	86.13	82.19	90.64	....
56	53.96	50.41	51.96	50.36	49.31	53.23	48.92
57	46.07	54.91	53.06	54.00	53.84	60.90	48.11
58	97.29	91.89	99.78	89.18	85.71	94.80	85.89
59	73.07	81.63	74.07	....	84.61	64.86	....
60	96.70	98.61	90.81	88.78	89.90	....	....

* Not in sagittal plane at occiput.

TABLE 1—Continued.

1	H, 22.	H, 23.	H, 24.	H, 25.	H, 26.	H, 27.	H, 28.
2							
3	4th ?	4th	4th	5th	5th	5th	1st
4	?	?	M	M	M	?	?
5							
6	160	176	154	166*	155	163	*160
7	161	176	158	165e	156	164	157
8	155	165	150		150	153	158½
9		138	144	147	146e	136	132
10		100?	104		104		99
11		87	70	81e		72	72
12		113	103	112e	110	95	95
13			81	89e		86	76
14	89; ——	86; 120	96e; ——		85; 112	89; 111e	86; 107
15		115, 115	122; 123		112; 112	111; 111	110; 110
16	480e	498	484		407	483	460
17		254; 244	224; 260		210; 257	233; 250	203; 257
18		330	345		327e	350	328
19	125; 131; ——	139; 107; 104	124; 121; 100	—; 96	108; 110; 108e		111; 117; 100
20		320; 307	345; 319		320; 296	318; 300	318; 287
21		30; 31	334; 304	37½; 33	—; 28	51; 28	36½; 30
22			129	141e		125	
23	106	104e	101	108e	101	105	89
24	104	105	95			95	79
25	21½		23	28		23	18
26	—; 68	108; 66	122; 72	125; 76	48; 29½e	—; 75	92; 54
27	46; 26	51; ——	50; 26	56; 27	38½e; 33	53e; 31	40; 22½
28	38½; 31½		138; 35½	41; 27	—; 38	46; 38	35; 31
29	55; ——	—; 38	49; 35	57; 26	—; ——	51; 34	38; 30
30		38; 23	38; 23	37; 23	39; ——	34; 25	32; 23
31			17			14	9
32		63; 49	59; 52	63; 49		61; 44	—; 37
33		13	15			31	5e
34		122e; 100	111; 105	110; 99			94½; 77
35		68; 37	50; 29	66; 33	—; 31		42½; 23
36		113	100	113			86
37		99	80	99		97½	80
38		105	95	105		97	85
39		120	108		99e	108	96
40		125	117		113e	114	106
41		120; 139	143; 146		120; 131e	134; 135	122; 124
42		132	140	134	124e		117
43		75	72?	72	66e	78	
44	197	201	201	108		206	172
45	184	193	196e	188	172e	195e	168
46		106; 93	90; 97	101; 93	94½; 83e	103; 93	82; 85
47	76e	84e	85e	84e		86e	88e
48		18½e	21½e	23		8e	9½e
49		284e; 38e	34e; 45e	35e; 46½e		29e; 26e	184e; 26½e
50		78, 46	91, 13	83, 09	89, 71	82, 92	84, 07
51		78, 97	90, 50		82, 69	81, 76	77, 70
52		79, 48	91, 01	88, 00		90, 32	82, 19
53		102, 85	128, 42				116, 45
54	65, 38	82, 85	75, 78			78, 94	68, 33
55			94, 57	88, 65			
56			55, 81	53, 19		69, 60	
57	56, 52		50, 60	49, 09	53, 12	45, 28	55, 55
58	81, 81		83, 42	86, 24	85, 71	96, 60	88, 57
59			71, 42	63, 15		62, 96	78, 94
60		94, 28	93, 68	91, 28		100, 51	94, 11

* Not strictly in sagittal plane.   † Not parallel to either border.

TABLE 1—Continued.

1	H. 29.	H. 32.*	H. 33.	H. 34.	H. 35.	H. 36.	H. 37.
2							
3	2nd	1th †	4th	6th	5th	6th	4th
4	?	M.	?	M.	?	F.	?
5							
6	119	† 148	164	169	† 157	† 160	167
7	151	148	166c	171	157	159	160
8	145	139	153 †	164	144	157	162
9	111	139c	145	143	135	135	143
10	95		111			100	...
11	73		81 †	84		74	...
12	97	110	106	105	102	104	114
13	78	80 †	83 †	85			92c
14	90; 113	85; 112	92c; —	92; 118		90; —	96c; 120 †
15	112; 112	121; 121	113; 113	119; 119	112; 112	117; 117	124; 124
16	463	448	495	499	461	472	492
17	208; 255	215; 233	230; 295	232c; 260c	207; 254		241; 251
18	323	338	361	361	333	343c	366c
19	114; 93; 117	120; 124; 94	133; 117; 111	137c; 113; 121	123; 115; 95		135; 116; 115c
20	333; 300	334; 311	327; 311	337; 313	319; 296	322; —	350; 326
21	35 †; 30		31; 132	34; 30 †		34; 28	—; 35
22	114	122	133				
23	93	99	108c	107c		101c	
24	90	92	90	98			
25	21	24c	30c		24		
26	98; 36	—; 64	—; 65c		116; 71		111c; 66
27	42; 22 ½	47; 24	45; 254	55; 24 ½	48; 25		47 ½; 26
28	35; 32	38c; 36				36÷c; 36 ½	38; 36
29	43; 33	48; —	53; 40	50; —	53; 37		55; 37
30			38; 24		38; 23		—; 25
31	11		21				16
32	61; 41	—; 44	64; 47	—; 49	—; 47	—; 51	61; —
33		6	21c	10c	23		25c
34	103; 70	110; 95				119; 92	113; 97
35	46; 30	58; 27	83; 32	63; 32	69; 30	57; 30	61; 35
36	86				105c		
37	80	92	96c	693	97 †		102 †
38	85	101c	95c	102	98	95c	95
39	95	113c	108c	111	109	106	106
40	102	113c	119 †	119	117	111	115
41	125; 131	144c; 148c	137; 137	43; 143	136; 138	—; 136	144; 143
42	132	130c	136	138	128	124	135c
43	60	64c		72		66	64
44	182c	188	202c		202c		202 †
45	175	181	193	194	181	182	190 †
46	81; 85	95; 61	97; 191	95; 84	95; 89	94; 81	—; 95c
47		85c	79c		77 ½		80
48	11c		114c	13c		12c	
49	20 ½; 28c		29 ½ ?; 30c	23 ½ ?; 31c		21 ÷ ?; 28c	
50	93, 37	91, 89	87, 34	83, 62	85, 92	84, 90	86, 14
51	82, 78	97, 29	82, 53	83, 62	86, 62		86, 74
52	84, 50		103, 22	89, 70		82, 35	
53	108, 88						
54	62, 22	69, 56	65, 65				
55	85, 96						
56	49, 12	52, 45	48, 87				
57	53, 17	51, 06	56, 66	44, 51	52, 08	59, 31	54, 73
58	91, 42	94, 73				100, 00	54, 73
59	77, 90		75, 47		69, 81		67, 67
60	94, 11	91, 08	101, 05	91, 17	98, 36		107, 36

*The conditions of Nos. 30 and 31 admitted of so few measurements that these are not tabulated.
†Not in sagittal plane at occiput.
'Taken at first permanent molars, the only molars erupted.
§General absorption of the alveoli. This is practically a basilo-subnasal line.

TABLE I—Continued.

1	H. 38.	H. 39.	H. 40.	H. 41.	H. 42.	H. 43.	H. 44.
2							
3	4th	4th	4th	4th	5th	4th	4th
4	?	F.	?	M.	?	?	?
5							
6			*158	161	*176		168
7		150	156	162	175		168
8		148	149	167	167		168
9	138	142	135	139	144		143
10	106	106	103			93	111
11	82		78	84e		80	
12	111	104e	105	114		107	108
13	88		81		81?	*6	
14		86; ---	91; 116e	92;	93; ---	82; 113	
15	123; 123	---; 117	113; 113	120; 120	116; 116		120; 120
16			405	402	500		403
17			211; 254	251; 261	241; 268		240; 253
18		329	328e	319e	361e		318e
19		118; 118; 93	114e; 116e; 99;	116; 127; 106e	125; 132; 104e		127; 117; 104e
20	330; ---	331; 305	312; 290	340; 311e	326; 311		333; 308
21	34; 31		354; 30		---; 27	36; 29	
22			125	196		133	
23			101	104	105	100	
24			93	98		92	
25						19	
26			114; 65	115; 64	123; 75	117; 71	---; 69e
27			45e; 23	48e; 26½	52; 27	52; 23	51e; 29
28			38; 36½	37½; 35½		36; 33	
29			30½; 38	50; 40		57; 40	50; ---
30			40; 19	40; 25	---; 27	38; 22	40; ---
31			16	17		19	
32			60; 45	64; 48		62; 46	---; 52
33			12	11		15½	13
34			108; 93	122; 98	16	115; 90	---; 116
35	69; 34	64; 29	53; 29	67; 35	71; 33	64; 33½	62; 32
36			90	115*	108e	106	
37			91	94	102	91	93
38			96e	97	109	92	106e
39			110	110	119	106	116
40			118	118	127	116	122
41	143e; 144		132e; 134	137; 139	140; 142		143; 143
42	132		127	116	133		
43	73		69		70	67	80
44			183	172	190		172
45				164e	188		170
46	---; 92		93; 83	99; 90	108e; 90	93; ---	94; 86
47			83?	84?	84?	87?	86?
48			14?			144?	
49			22½?; 30?			26?; 344?	
50		94. 66	86. 53	91. 97	82. 28		85. 11
51			84. 61	84. 56	80. 00		
52	91. 17		84. 50			80. 55	
53			122. 58	117. 34		127. 17	
54			69. 89	65. 30		77. 17	
55			92. 68	84. 55		87. 97	
56			52. 84	47. 65		53. 38	
57			51. 11	53. 12	51. 92	44. 23	56. 86
58			96. 05	94. 66		91. 66	
59			76. 00	80. 00		71. 05	
60			94. 79	96. 80	93. 57	98. 91	93. 00

* Not in sagittal plane at occiput.

TABLE I—Continued.

1	H. 45.	H. 46.	H. 47.	H. 49.	H. 50.	H. 51.	H. 52.
2							
3	14th	3d	4th	4th	2d	2d	4th
4	F.	?	?	?	?	?	?
5							
6	161	146	*153		150	145	*160
7	163	166e	159		150	148	159
8	154	139	144		145	141	161
9	148	145	147		136	139	137
10			98		92	195	103
11		81	88		79		
12		104	109	108	102	96	111
13	88				79	77	
14	94; —	95; 114e	93; —		88; 108e	87; 114	91; 119
15	116; 116	113; 113	125; 125		111; 111	112; 112	118; 118
16	497	468	448c		451	456	478
17	221; 276	209; 259	290e; 263		303; 248	194; 262	259; 279
18	333		344		324	339e	342
19	120; 167e; 108e	—; 100; 108	123; 124; 97	120; 120; —	119; 107; 98	115; 110; 105	120; 114; 108
20	330; 310	322?; 300	350; 324		307; 290	318; 306	322; 302
21		32e; 33e	32; 31		384; 284	—; 30	
22	120						
23	104	104			96	91	100
24	96	91			92	85	
25				27½		30	
26	109; 66			—; 71	—; 60	—; 57	—; 71
27	47; 28½			48; 28½†	45; 24	42½; 26	53; 24
28	38; 33½			36; 34		33; 30	
29	53; 38	40; 38		50; 38	42; 32½	45; —	—; 36
30	42; 25			41; —	—; 27		36; —
31	14			16	11		
32	65; 54			65; —	59	57; 42	67; —
33	17	20			10	10	
34	122; 99		111; 97				
35	60; 54	54; 39	64; 32	60; 38			
36	110						
37	99	95			82	81	
38	101		102		89	87	
39	108	100e	111		101	68	
40	119	108	120		111	106	
41	135; 136	125; 126	144; 144		131; 132	131; 132	
42			136		123	125	
43		50	60?		68e	56e	
44	199	188			176	182	173
45	190	178			168	175	161
46	100; 90	91; 81	91; 81		83; 81	82; 77	97?; 78?
47	82°	82½°		85°	85°	86°	85°
48		14½°			14½°		13°
49		28°; 34½°			21½°; 35°		21°; 30°
50	90, 79	97, 97	92, 45		90, 66	93, 91	85, 62
51	82, 82	83, 10	90, 56		87, 33	88, 51	86, 25
52		103, 12	96, 87		74, 02		
53	113, 54						
54	68, 75				65, 21	67, 05	
55	84, 49						45, 28
56	51, 16						
57	32, 13			59, 37	55, 83	47, 05	
58	88, 15			94, 44		93, 90	
59	71, 69	77, 35		76, 00	76, 19		
60	98, 01				92, 13	93, 10	

* Not in sagittal plane at occiput.
† This is the true maximum circumference; the circumference at the maximum occipital point (which coincides with the obelion) is 471.
‡ Taken at the first permanent molar, the second not being erupted.

No. 48 is not **included** in this table, because its condition did not admit of many measurements.

Table I—Concluded.

1	H. 53.	H. 54.	H. 55.	H. 56.	H. 57.
2					
3	5th ?	4th	5th	5th	4th
4	?	?	?	?	F.
5	1,120cc				
6	*149	145	188	*159	*132
7	148	148	188	158	150
8	145	146	186	152	146c
9	136	134	148	153	142
10	100	101c	107		
11	77		74		
12	98	97	104	85c; 115c	101
13	76?				
14	83; 115	89; —	100; —	92; —	91
15	115; 113	113; 113	127; 127	113; 113	121; 121
16	450	450	530c	476	464
17	201; 249	206; 244	275; 285	220; 247	219; 245
18	340	330			225
19	122; 115; 103	117; 113; 100		120; 113; —	120; 111; 94
20	330; 303	320; 301	352; 307	323; 302	322; 302
21	314; 284				384; —
22				130	
23	96	101	113	102	102
24					
25				24	
26		111; 67		106; 64	104; 62
27		48; —		46; 28½	45; 27½
28	34½; 32			38; 35	37; 32½
29		—; 36		52; 38	52; 40
30		—; 24		38; —	40; 27
31					15
32				—; 48	63; 46
33		25		15	20c
34		108; 56		120c; 87	
35		61; 26		56; 32	53; 30
36					108
37					40
38	92				89
39	105				28
40	118				109
41	137; 137			132c; —	128; 129
42	123				123
43	61?				63
44		182		186	183
45	167e	170		178	172
46	—; 76	91; 90			91; 85
47		84"			82"
48	94?			84"	114?
49	20?; 264?				22?; 30½?
50	91.80	90.54	78.72	90.50	94.66
51	92.56	87.16		83.54	85.33
52	90.47				
53					
54					
55				81.55	
56				49.23	
57				53.26	61.11
58	92.75			92.16	87.83
59				73.07	76.92
60					101.12

* Not in sagittal plane at occiput.

TABLE II.—*Ordination of 48 cephalic indices.—Salado.*

No. of skull	Index.	No. of skull	Index.	No. of skull	Index.
1	H. 23	78. 40	17	H. 40	86. 53	33	H. 50	90. 36
2	H. 55	78. 72	18	H. 19	86. 54	34	H. 45	90. 79
3	H. 12	78. 97	19	H. 33	87. 34	35	H. 24	91. 43
4	H. 42	82. 38	20	H. 9	87. 41	36	H. 32	91. 89
5	H. 37	82. 92	21	H. 5	87. 50	37	H. 53	91. 89
6	H. 34	83. 62	22	H. 13	87. 58	38	H. 41	91. 97
7	H. 28	84. 07	23	H. 11	87. 89	39	H. 21	92. 45
8	H. 15	84. 27	24	H. 8	88. 07	40	H. 47	92. 45
9	H. 16	84. 27	25	H. 17	88. 41	41	H. 29	93. 37
10	H. 26	84. 90	26	H. 14	88. 55	42	H. 6	93. 71
11	H. 44	85. 11	27	H. 25	89. 09	43	H. 51	93. 91
12	H. 52	85. 62	28	H. 26	89. 74	44	H. 39	94. 66
13	H. 18	85. 79	29	H. 2	90. 00	45	H. 57	94. 66
14	H. 7	85. 80	30	H. 10	90. 50	46	H. 4	96. 00
15	H. 35	85. 92	31	H. 56	90. 90	47	H. 3	96. 17
16	H. 37	86. 14	32	H. 54	90. 54	48	H. 46	97. 97

Variation, 19.57. Theoretical mean of variation, 88.19. Skull nearest to mean, H. 8. Average, 88.47.

TABLE III.—*Seriation of 48 cephalic indices.—Salado.*

	Index.	Number of skulls.		Index.	Number of skulls.
1	77 to 78	1	12	88	3
2	78	2	13	89	2
3	79	0	14	90	6
4	80	0	15	91	4
5	81	0	16	92	2
6	82	2	17	93	3
7	83	1	18	94	2
8	84	4	19	95	0
9	85	5	20	96	2
10	86	3	21	97 to 98	1
11	87	5			

Maximum of frequency, 90.

TABLE IV.—*Ordination of 16 cephalic indices of apparently normal skulls.—Salado.*

No.	Index.	No.	Index.
1	H. 23	78. 40	9	H. 40	86. 53
2	H. 12	78. 97	10	H. 19	86. 54
3	H. 34	83. 62	11	H. 25	89. 09
4	H. 15	84. 27	12	H. 26	89. 74
5	H. 36	84. 90	13	H. 54	90. 54
6	H. 44	85. 11	14	H. 21	92. 45
7	H. 18	85. 79	15	H. 39	94. 66
8	H. 7	85. 80	16	H. 57	94. 66

Variation, 16.20. Theoretical mean of variation, 86.53. Skulls nearest to mean, H. 7 and H. 40. Average, 86.94.

No. H. 55, an apparently normal skull, of which the vault only is preserved, has a normal index of 78.72, but it is aberrant as regards the rest of the group by reason of its much greater size and different configuration.

No. H. 23, having lowest cephalic index, has next to lowest vertico-transverse index. The lowest vertico-transverse index is in the skull of a child.

TABLE V.—*Seriation of 16 cephalic indices of apparently normal skulls.—Salado.*

	Index.	Number of skulls.		Index.	Number of skulls.
1	78 to 79	2	10	87	0
2	79	0	11	88	0
3	80	0	12	89	2
4	81	0	13	90	1
5	82	0	14	91	0
6	83	1	15	92	1
7	84	2	16	93	0
8	85 to 86	3	17	94 to 95	2
9	86	2			

Maximum of frequency, 85 to 86.

TABLE VI.—*Of the cephalic indices of all the skulls.—Salado.*

	48	Per cent.
Number of indices............................		
Number below 80.00........................	3	6.25
Number from 80.00 to 89.99.................	25	52.09
Number from 90.00 up......................	20	41.66
	48	100.00

Minimum index, 77.65. Maximum index, 97.97.

TABLE VII.—*Ordination of 47 length-breadth indices (German).—Salado.*

Number of skull.	Index.	Number of skull.	Index.	Number of skull.	Index.
1	H. 23	78.40	17	H. 35	85.98	33	H. 21	91.87
2	H. 55	78.72	18	H. 37	86.60	34	H. 32	91.89
3	H. 12	78.97	19	H. 9	86.84	35	H. 45	91.92
4	H. 42	81.81	20	H. 5	87.50	36	H. 10	92.25
5	H. 28	82.50	21	H. 11	87.89	37	H. 54	92.41
6	H. 27	83.43	22	H. 13	88.15	38	H. 41	92.54
7	H. 36	84.37	23	H. 33	88.41	39	H. 57	93.42
8	H. 19	84.57	24	H. 14	88.55	40	H. 24	93.50
9	H. 34	84.61	25	H. 25	88.55	41	H. 6	93.71
10	H. 15	84.81	26	H. 8	89.44	42	H. 29	94.63
11	H. 16	84.81	27	H. 56	89.92	43	H. 51	95.86
12	H. 44	85.11	28	H. 26	90.32	44	H. 47	96.07
13	H. 40	85.44	29	H. 17	90.62	45	H. 3	96.17
14	H. 52	85.62	30	H. 50	90.66	46	H. 4	98.65
15	H. 38	85.79	31	H. 53	91.27	47	H. 46	99.31
16	H. 7	85.80	32	H. 2	91.72			

Variation, 20.91. Theoretical mean of variation, 88.86. Skulls nearest to mean, H. 14 and H. 25. Average, 88.75.

TABLE VIII.—*Ordination of 38 vertico-longitudinal indices.—Salado.*

Number of skull.	Index.	Number of skull.	Index.	Number of skull.	Index.
1	H. 28	78.79	14	H. 56	83.54	27	H. 37	86.74
2	H. 15	79.87	15	H. 34	83.62	28	H. 54	87.16
3	H. 42	80.00	16	H. 41	84.56	29	H. 50	87.33
4	H. 27	81.70	17	H. 40	84.61	30	H. 51	88.51
5	H. 33	82.53	18	H. 11	84.71	31	H. 3	88.89
6	H. 26	82.69	19	H. 44	85.11	32	H. 9	89.40
7	H. 29	82.78	20	H. 4	85.33	33	H. 2	90.00
8	H. 45	82.82	21	H. 57	85.33	34	H. 24	90.50
9	H. 17	82.92	22	H. 19	85.38	35	H. 47	90.56
10	H. 46	83.10	23	H. 52	86.25	36	H. 13	91.50
11	H. 14	83.13	24	H. 1	86.30	37	H. 10	91.77
12	H. 5	83.33	25	H. 18	86.39	38	H. 53	92.86
13	H. 7	83.33	26	H. 35	86.62			

Variation, 13.59. Theoretical mean of variation, 85.76. Skull nearest to mean, H. 19. Average, 83.87.

H. 28, a child's skull, with index of 77.70, and H. 32, aberrant in size and form (see § 11), with index of 97.29, are excluded from the above ordination.

TABLE IX.—*Seriation of 39 vertico-longitudinal indices.—Salado.*

Index.	Number of skulls.	Index.	Number of skulls.
78 to 79	1	88	2
79	1	89	1
80	1	90	3
81	1	91	2
82	5	92	1
83	6	93	0
84	3	94	0
85	4	95	0
86	5	96	0
87	2	97 to 98	1

Maximum of frequency, 83; 97 to 98, one skull, not included in the ordination.

TABLE X.—*Ordination of 11 vertico-longitudinal indices of apparently normal skulls.—Salado.*

	Number of skull.	Index.		Number of skull.	Index.
1	H. 23	78.97	7	H. 44	85.11
2	H. 15	79.87	8	H. 57	85.33
3	H. 26	82.09	9	H. 19	85.38
4	H. 7	83.33	10	H. 38	86.36
5	H. 31	83.62	11	H. 54	87.16
6	H. 40	84.61			

TABLE XI.—*Seriation of 11 vertico-longitudinal indices of apparently normal skulls.—Salado.*

Index.	Number of skulls.	Index.	Number of skulls.
78 to 79	1	83	2
79	1	84	1
80	0	85	3
81	0	86	1
82	1	87 to 88	1

TABLE XII.—*Ordination of all the skulls in which the vertico-longitudinal index exceeds the cephalic, showing excess in per cent of greatest length.—Salado.*

	No.	Excess.		No.	Excess.
1	H. 18	0.50	6	H. 35	0.70
2	H. 23	0.57	7	H. 16	1.20
3	H. 37	0.60	8	H. 9	1.99
4	H. 52	0.63	9	H. 13	3.92
5	H. 53	0.67	10	H. 32	5.40

Variation, 4.90. Theoretical mean of variation, 2.95. Skulls nearest to mean, H. 9, and H. 13. Skulls H. 18 and H. 23 are from the apparently normal group.

TABLE XIII.—*Seriation of all the skulls in which the vertico-longitudinal index exceeds the cephalic, showing excess in per cent of greatest length.—Salado.*

	Excess.	Number of skulls.		Excess.	Number of skulls.
1	0 to 1	6	4	3	1
2	1	2	5	4	0
3	2	0	6	5 to 6	1

TABLE XIV.—*Ordination of vertico-transverse indices.—Salado.*

No. of skull.	Width.	Height.	Index.	No. of skull.	Width.	Height.	Index.
1	H. 46	14.5	12.3	84.82	21	H. 49	13.5	13.2	97.77
2	H. 4	14.5	12.8	88.27	22	H. 47	14.7	14.4	97.95
3	H. 29	14.1	12.5	88.65	23	H. 27	13.6	13.4	98.52
4	H. 57	14.2	12.8	90.14	24	H. 19	14.8	14.6	98.64
5	H. 45	14.8	13.5	91.21	25	H. 24	14.4	14.3	99.30
6	H. 41	14.9	13.7	91.94	26	H. 2	14.4	14.4	100.00
7	H. 36	14.0	12.9	92.14	27	H. 34	14.3	14.3	100.00
8	H. 56	14.3	13.2	92.30	28	H. 44	14.3	14.3	100.00
9	H. 28	13.2	12.2	92.42	29	H. 18	14.5	14.6	100.68
10	H. 17	14.5	13.6	93.79	30	H. 37	14.3	14.4	100.69
11	H. 14	14.7	13.8	93.87	31	H. 23	13.8	13.9	100.72
12	H. 51	13.9	13.1	94.24	32	H. 52	13.7	13.8	100.72
13	H. 33	14.5	13.7	94.48	33	H. 53	13.6	13.7	100.73
14	H. 15	13.4	12.7	94.77	34	H. 35	13.5	13.6	100.74
15	H. 5	13.7	14.0	95.23	35	H. 10	14.3	14.5	101.39
16	H. 54	13.4	12.9	96.26	36	H. 9	13.2	13.5	102.27
17	H. 50	13.6	13.1	96.32	37	H. 38	13.8	14.3	103.62
18	H. 11	13.8	13.3	96.37	38	H. 13	13.4	14.0	104.47
19	H. 7	13.9	13.5	97.12	39	H. 32	13.6	14.4	105.88
20	H. 42	14.4	14.0	97.22					

TABLE XV.—*Of the vertico-transverse indices of all the skulls.—Salado.*

		Per cent.
Number of indices	40	
Number below 80.00	3	7.5
Number from 80.00 to 89.99	30	75.0
Number from 90.00 up	7	17.5
	40	100.0

Minimum index, 77.70; maximum index, 97.29. Neither of these two indices come in the normal series, as the minimum is that of a child, the maximum aberrant. They therefore do not appear in the ordination.

TABLE XVI.—*Ordination of the apparently normal skulls, with reference to the differences between their respective cephalic and vertico-transverse indices, expressed in per cent of the greatest length.—Salado.*

[The sign + indicates that the cephalic index is greater than the vertico-transverse. The sign — indicates that the vertico-transverse index is greater than the cephalic.]

	No.	Difference.		No.	Difference.
1	H. 23	—.57	7	H. 7	+2.47
2	H. 18	—.50	8	H. 54	+5.38
3	H. 34	.00	9	H. 15	+4.40
4	H. 44	.00	10	H. 26	+7.05
5	H. 19	+1.16	11	H. 57	+9.33
6	H. 40	+1.92			

Variation, 9.90. Theoretical mean of variation, 4.38. Skull nearest to mean, H. 15. This skull then shows what may be arithmetically regarded as a typical relation of vertico-transverse and cephalic indices.

TABLE XVII.—*Ordination of mixed indices.—Salado.*

	No. of skull	Vertico-transverse index.	Vertico-longitudinal index.	Mixed index.
1	H. 46	84. 82	83. 10	83. 96
2	H. 20	88. 65	82. 78	85. 71
3	H. 4	88. 27	85. 33	86. 80
4	H. 45	91. 21	82. 82	87. 01
5	H. 26	92. 14	82. 69	87. 41
6	H. 15	93. 97	79. 87	87. 42
7	H. 37	90. 14	85. 33	87. 73
8	H. 36	92. 30	83. 54	87. 92
9	H. 41	91. 94	84. 56	88. 25
10	H. 17	93. 79	82. 92	88. 35
11	H. 33	94. 48	82. 51	88. 50
12	H. 11	93. 87	83. 13	88. 50
13	H. 42	97. 22	80. 00	88. 61
14	H. 5	95. 23	83. 33	89. 28
15	H. 23	100. 72	78. 79	89. 75
16	H. 27	98. 52	81. 70	90. 11
17	H. 7	97. 12	83. 33	90. 22
18	H. 11	96. 37	84. 71	90. 54
19	H. 40	97. 77	84. 61	91. 19
20	H. 54	94. 24	88. 51	91. 37
21	H. 54	96. 26	87. 16	91. 71
22	H. 34	100. 00	83. 62	91. 81
23	H. 50	96. 32	87. 33	91. 82
24	H. 19	98. 64	85. 38	92. 01
25	H. 44	100. 00	85. 11	92. 55
26	H. 32	100. 72	86. 25	93. 48
27	H. 18	100. 68	86. 39	93. 53
28	H. 35	100. 74	86. 62	93. 68
29	H. 37	100. 69	86. 74	93. 71
30	H. 17	97. 95	90. 56	94. 25
31	H. 24	99. 30	90. 50	94. 90
32	H. 2	100. 00	90. 00	95. 00
33	H. 9	102. 27	89. 40	95. 83
34	H. 10	101. 39	91. 77	96. 58
35	H. 53	100. 73	92. 56	96. 64
36	H. 13	104. 47	91. 50	97. 98
	Average..	96. 46	85. 40	90. 94

TABLE XVIII.—*Ordination of 29 angles of Daubenton.—Salado.*

	Number of skull.	Angle.			Number of skull.	Angle.	
		°	′			°	′
1	H. 45	4	30	16	H. 19	14	00
2	H. 16	7	30	17	H. 40	14	00
3	H. 27	8	00	18	H. 43	14	30
4	H. 28	9	30	19	H. 46	14	30
5	H. 53	9	30	20	H. 50	14	30
6	H. 6	10	00	21	H. 3	15	00
7	H. 8	10	50	22	H. 14	15	30
8	H. 5	11	00	23	H. 7	16	00
9	H. 29	11	00	24	H. 17	17	00
10	H. 33	11	30	25	H. 10	18	30
11	H. 57	11	30	26	H. 23	18	30
12	H. 36	12	00	27	H. 18	20	00
13	H. 20	12	30	28	H. 24	21	30
14	H. 34	13	00	29	H. 25	23	00
15	H. 52	13	00				

Variation, 18° 30′. Theoretical **mean**, 13° 45′. Skulls nearest to mean, H. 19 and H. 40. Average, 13° 30′

TABLE XIX.—*Seriation of 29 angles of Daubenton.—Salado.*

	Angle.	Number of skulls.		Angle.	Number of skulls.
1	4° to 5°	1	11	14°	5
2	5	0	12	15	2
3	6	0	13	16	1
4	7	1	14	17	1
5	8	1	15	18	2
6	9	2	16	19	0
7	10	2	17	20	1
8	11	4	18	21	1
9	12	2	19	22	0
10	13	2	20	23 to 24	1

Maximum of frequency, 14°.

TABLE XX.—*Ordination of 29 occipital angles.— Salado.*

No. of skull.	Angle.		No. of skull.	Angle.
	° ′			° ′
1	H. 15	13 30	16	H. 3	24 00
2	H. 28	18 30	17	H. 52	24 00
3	H. 6	19 00	18	H. 50	24 30
4	H. 16	19 00	19	H. 44	25 00
5	H. 5	20 00	20	H. 39	25 30
6	H. 27	20 00	21	H. 43	26 00
7	H. 29	20 00	22	H. 7	27 30
8	H. 53	20 00	23	H. 46	28 00
9	H. 8	21 30	24	H. 23	28 30
10	H. 36	21 30	25	H. 17	29 00
11	H. 57	22 00	26	H. 10	30 00
12	H. 49	22 30	27	H. 18	30 30
13	H. 20	23 00	28	H. 24	34 00
14	H. 33	23 30	29	H. 25	35 00
15	H. 34	23 30			

Variation, 21° 30′. Theoretical mean, 24° 15′. Skulls nearest to mean, H. 3, H. 52, H. 50.
Average, 24° 6′.

TABLE XXI.—*Seriation of 29 occipital angles.—Salado.*

	Angle.	Number of skulls.		Angle.	Number of skulls.
1	13° to 14°	1	13	25°	2
2	14	0	14	26	1
3	15	0	15	27	1
4	16	0	16	28	2
5	17	0	17	29	1
6	18	1	18	30	2
7	19	2	19	31	0
8	20	4	20	32	0
9	21	2	21	33	0
10	22	2	22	34	1
11	23	3	23	35 to 36	1
12	24	3			

Maximum of frequency, 20°

TABLE XXII.—*Ordination of 29 basilar angles.—Salado.*

No. of skull.	Angle.	No. of skull.	Angle.
1	H.15	18 00	16	H.34	31 00
2	H.6	24 00	17	H.3	31 30
3	H.16	25 00	18	H.19	32 30
4	H.5	26 00	19	H.11	33 30
5	H.27	28 00	20	H.43	34 30
6	H.28	26 30	21	H.46	34 30
7	H.53	26 30	22	H.50	35 00
8	H.29	28 00	23	H.7	36 30
9	H.36	28 00	24	H.10	28 00
10	H.8	28 30	25	H.24	38 00
11	H.33	30 00	26	H.47	38 30
12	H.40	30 00	27	H.18	40 00
13	H.52	30 00	28	H.21	45 00
14	H.29	30 30	29	H.25	46 30
15	H.37	30 30			

Variation, 28° 30′. Theoretical mean, 32° 15′. Skull nearest to mean, H.19. Average, 31° 48′.

TABLE XXIII.—*Seriation of 29 basilar angles.—Salado.*

Angle.	Number of skulls.	Angle.	Number of skulls.
1	18° to 19°	1	16	33°	1
2	19	0	17	34	2
3	20	0	18	35	1
4	21	0	19	36	1
5	22	0	20	37	0
6	23	0	21	38	3
7	24	1	22	39	0
8	25	1	23	40	1
9	26	4	24	41	0
10	27	0	25	42	0
11	28	3	26	43	0
12	29	0	27	44	0
13	30	5	28	45	1
14	31	2	29	46 to 47	1
15	32	1			

Maximum of frequency, 30°.

TABLE XXIV.—*Average cranial capacity of 12 small series of skulls in the general collection of the Army Medical Museum.*

Races or tribes.	Total number of skulls.	Average capacity.
Sandwich Islanders	6	1491
Mongolians (2 Japanese, 2 Chinese)	4	1465
Siouan (4 Sioux, 4 Poncas, 2 Minnetarees)	10	1463
New Zealanders	4	1453
American negroes	6	1383
North American Indians (11 tribes, exclusive of Saladoans)	47	1374
Pah-Utes	7	1367
Eskimos (4 Alaska, 6 Greenland)	10	1357
Apaches	6	1331
Ancient Californians	10	1323
Navajos	6	1345
Peruvians (4 artificially elongated, 3 with antero-posterior compression, 2 normal, 1 plagio-cephalic)	10	1295

TABLE XXV.—*Length in millimetres of 28 pteria.—Salado.*

No. of skull.	Right side.	Left side.	No. of skull.	Right side.	Left side.
H. 7	....	11	H. 19	10	12
H. 10	10	9	H. 22	....	9
H. 11	....	9	H. 24	....	8
H. 12	....	16	H. 28	13	12
H. 13	5	....	H. 29	13	16
H. 14	....	12	H. 38	10	....
H. 15	....	3	H. 40	....	6
H. 16	11	12	H. 42	10	....
H. 17	20	18	H. 50	....	14
H. 18	15	10	H. 51	16	16

TABLE XXVI.—*Ordination of 19 facial indices according to Virchow.—Salado.*

	No. of skull.	Index.		No. of skull.	Index.
1	H. 23	102.85	11	H. 11	118.00
2	H. 16	106.45	12	H. 18	120.40
3	H. 29	108.88	13	H. 19	121.21
4	H. 17	109.00	14	H. 40	122.58
5	H. 5	110.67	15	H. 43	127.17
6	H. 1	111.65	16	H. 14	127.55
7	H. 4	112.24	17	H. 24	128.42
8	H. 45	113.54	18	H. 15	129.67
9	H. 28	116.45	19	H. 20	131.25
10	H. 41	117.34			

Variation, 28.40. Theoretical mean of variation, 117.05. Skull nearest to mean and median of ordination, H. 41. Average, 117.64.

TABLE XXVII.—*Seriation of 19 facial indices according to Virchow.—Salado.*

Index.	Number of skulls.	Index.	Number of skulls.	Index.	Number of skulls.
102 to 103	1	112	1	122	1
103	0	113	1	123	0
104	0	114	0	124	0
105	0	115	0	125	0
106	1	116	1	126	0
107	0	117	1	127	2
108	1	118	1	128	1
109	1	119	0	129	1
110	1	120	1	130	0
111	1	121	1	131 to 132	1

Maximum of frequency, 127. The seriation is so incoherent that the discussion of variation as dependent on it has little significance.

TABLE XXVIII.—*Ordination of 34 upper facial indices according to Virchow.—Salado.*

	No. of skull.	Index.		No. of skull.	Index.		No. of skull.	Index.
1	H. 29	62.22	13	H. 51	67.65	25	H. 19	72.72
2	H. 23	62.85	14	H. 4	67.34	26	H. 8	72.92
3	H. 50	65.21	15	H. 9	68.51	27	H. 3	73.71
4	H. 41	65.39	16	H. 28	68.35	28	H. 15	74.72
5	H. 22	65.58	17	H. 2	68.42	29	H. 21	75.78
6	H. 16	65.59	18	H. 45	68.75	30	H. 10	76.34
7	H. 33	65.65	19	H. 6	69.32	31	H. 20	77.08
8	H. 7	66.00	20	H. 32	69.56	32	H. 43	77.17
9	H. 17	66.00	21	H. 40	69.89	33	H. 27	78.94
10	H. 5	66.01	22	H. 18	70.46	34	H. 14	79.59
11	H. 21	66.66	23	H. 11	71.00			
12	H. 13	67.00	24	H. 1	72.72			

Variation, 17.37. Theoretical mean of variation, 70.90. Skull nearest to mean, H. 11. Average, 69.82.

TABLE XXIX.—*Seriation of 31 upper facial indices according to Virchow.—Salado.*

Index	No. of skulls.	Index.	No. of skulls.
62 to 63	2	71	1
63	6	72	3
64	6	73	1
65	5	74	1
66	4	75	1
67	3	76	1
68	4	77	2
69	3	78	1
70	1	79 to 80	1

Maximum of frequency, 65.

TABLE XXX.—*Ordination of 17 facial indices according to Kollmann.—Salado.*

	No. of skull.	Index.		No. of skull.	Index.
1	H. 56	81.53	10	H. 43	87.97
2	H. 16	81.81	11	H. 25	88.65
3	H. 19	82.19	12	H. 20	90.64
4	H. 45	84.49	13	H. 11	91.47
5	H. 44	84.55	14	H. 40	92.68
6	H. 17	85.86	15	H. 15	93.65
7	H. 29	85.96	16	H. 24	94.57
8	H. 18	86.13	17	H. 14	97.65
9	H. 1	86.50			

Variation, 16.12. Theoretical mean of variation, 89.59. Skull nearest to mean, H. 25. Average, 88.01.

TABLE XXXI.—*Seriation of 17 facial indices according to Kollmann.—Salado.*

	Index.	Number of skulls.		Index.	Number of skulls.
1	81 to 82	2	10	90	1
2	82	1	11	91	1
3	83	0	12	92	1
4	84	2	13	93	1
5	85	2	14	94	1
6	86	2	15	95	0
7	87	1	16	96	0
8	88	1	17	97 to 98	1
9	89	0			

Maximum of frequency, 84, 85, and 86.

TABLE XXXII.—*Ordination of 27 upper facial indices according to Kollmann.—Salado.*

	No. of skull.	Index.		No. of skull.	Index.
1	H. 41	47.05	15	H. 13	52.75
2	H. 33	48.87	16	H. 40	52.84
3	H. 21	48.92	17	H. 25	53.19
4	H. 29	49.12	18	H. 20	53.23
5	H. 56	49.25	19	H. 43	53.38
6	H. 19	49.31	20	H. 1	53.96
7	H. 18	50.36	21	H. 15	53.96
8	H. 7	50.38	22	H. 9	54.76
9	H. 8	50.38	23	H. 11	55.03
10	H. 16	50.41	24	H. 24	55.81
11	H. 45	51.15	25	H. 39	56.34
12	H. 5	51.43	26	H. 27	60.00
13	H. 17	51.86	27	H. 14	60.93
14	H. 32	52.45			

Variation, 13.88.   Theoretical mean of variation, 53.99.   Skulls nearest to mean, H. 1 and H. 15.   Average, 52.48.

Excluding H. 27 and H. 14 the variation is 9.27, the mean 51.69, the skull nearest the mean H. 3, and the average 51.85.

TABLE XXXIII.—*Seriation of 27 upper facial indices according to Kollmann.—Salado.*

	Index	Number of skulls.		Index.	Number of skulls.
1	47 to 48	1	8	54	1
2	48	2	9	55	2
3	49	3	10	56	1
4	50	4	11	57	0
5	51	3	12	58	0
6	52	3	13	59	0
7	53	5	14	60 to 61	2

Maximum of frequency, 53.

TABLE XXXIV.—*Ordination of 44 German profile angles.—Salado.*

No. of skull.	Angle.		No. of skull.	Angle.		No. of skull.	Angle.
		° ′			° ′			° ′
1	H. 22	76 00	16	H. 46	82 30	31	H. 32	85 00
2	H. 4	77 00	17	H. 24	83 00	32	H. 50	85 00
3	H. 35	77 30	18	H. 46	83 00	33	H. 52	85 00
4	H. 2	78 00	19	H. 49	83 00	34	H. 6	86 00
5	H. 9	78 00	20	H. 7	84 00	35	H. 14	86 00
6	H. 33	78 00	21	H. 15	84 00	36	H. 51	86 00
7	H. 1	79 30	22	H. 16	84 00	37	H. 3	86 30
8	H. 10	80 00	23	H. 23	84 00	38	H. 44	86 30
9	H. 11	80 00	24	H. 25	84 00	39	H. 5	87 00
10	H. 27	80 00	25	H. 41	84 00	40	H. 43	87 00
11	H. 37	80 00	26	H. 42	84 00	41	H. 18	88 00
12	H. 13	81 00	27	H. 54	84 00	42	H. 28	88 00
13	H. 45	82 00	28	H. 56	84 00	43	H. 17	88 30
14	H. 57	82 00	29	H. 8	84 30	44	H. 19	89 00
15	H. 20	82 30	30	H. 21	85 00			

Variation, 13°.   Theoretical mean, 82° 30′.   Skulls nearest to mean, H. 20 and H. 46.   Average, 83° 25′.   Skulls nearest to average, H. 24, H. 40, H. 49.

TABLE XXXV.—*Seriation of 44 German profile angles.—Salado.*

	Angle.	Number of skulls.		Angle.	Number of skulls.
1	76° to 77°	1	8	83°	3
2	77	2	9	84	10
3	78	2	10	85	4
4	79	2	11	86	5
5	80	4	12	87	2
6	81	1	13	88	3
7	82	4	14	89 to 90	1

Maximum of frequency, 84°.

TABLE XXXVI.—*Ordination of 39 gnathic indices.—Salado.*

No. of skull.	Index.	No. of skull.	Index.	No. of skull.	Index.
1	H. 18	88.78	14	H. 24	93.68	27	H. 7	97.09
2	H. 19	89.90	15	H. 13	94.00	28	H. 45	98.01
3	H. 30	90.00	16	H. 28	94.11	29	H. 14	98.01
4	H. 17	90.81	17	H. 29	94.11	30	H. 35	98.36
5	H. 32	91.08	18	H. 22	94.28	31	H. 43	98.91
6	H. 31	91.17	19	H. 25	94.28	32	H. 11	100.00
7	H. 9	91.83	20	H. 40	94.79	33	H. 27	100.51
8	H. 3	92.07	21	H. 6	95.45	34	H. 1	101.01
9	H. 50	92.43	22	H. 8	95.91	35	H. 53	101.05
10	H. 44	93.00	23	H. 10	96.00	36	H. 57	101.12
11	H. 51	93.10	24	H. 5	96.11	37	H. 2	102.06
12	H. 42	93.57	25	H. 15	96.70	38	H. 37	107.36
13	H. 16	93.61	26	H. 41	96.90	39	H. 4	110.11

Variation, 21.33.   Theoretical mean, 90.44.   Skull nearest mean, H. 11.   **Average, 95.92.**
Skulls H. 7, H. 15, H. 18, H. 19, H. 23, H. 25, H. 34, H. 40, **H. 44, and H. 57 (10 in all)** are apparently normal. Their average index is 94.10.

TABLE XXXVII.—*Seriation of 39 gnathic indices.—Salado.*

Index.	Number of skulls.	Index.	Number of skulls.
88 to 89	1	100	2
89	1	101	3
90	2	102	1
91	3	103	0
92	2	104	0
93	5	105	0
94	6	106	0
95	2	107	1
96	4	108	0
97	1	109	0
98	4	110 to 111	1
99	0		

Maximum of frequency, 94.

TABLE XXXVIII.—*Seriation of 37 gnathic indices (aberrant figures excluded).—Salado.*

Index	Number of skulls.	Index.	Number of skulls.
88 to 89	1	96	4
89	1	97	1
90	2	98	4
91	3	99	0
92	2	100	2
93	5	101	3
94	6	102	1
95	2		

Variation, 13.28.   Theoretical mean, 95.42.   Skull nearest mean, H. 6.   Average, 95.20.

TABLE XXXIX.—*List of indices and angles of alveolo-subnasal prognathism.—Nalado.*

No. of skull.	Vertical measurement.	Horizontal measurement.	Index.	Angle.	No. of skull.	Vertical measurement.	Horizontal measurement.	Index.	Angle.
	mm.	mm.		°		mm.	mm.		°
H. 2	15	7.5	50.00	63½	H. 23	15	5	33.33	72
H. 3	19	4	21.05	78	H. 24	16	5	31.25	73
H. 4	15	8	53.33	62	H. 25	22	6	27.27	75
H. 5	21	9	42.85	67	H. 33	21	6	28.57	74½
H. 7	14	6.5	46.42	66	H. 35	20	9	45.00	66
H. 10	20	6.5	32.50	72	H. 37	16	9	56.25	62
H. 11	19	7.5	39.47	68	H. 42	22	7	31.81	72½
H. 14	18	8	44.44	66	H. 43	21	3	14.28	82
H. 15	17	10	58.82	60	H. 45	16	5	31.25	73
H. 16	16	5	31.25	73	H. 46	15	7	46.66	66
H. 17	17	3	17.64	79½	H. 54	17	7	41.17	68
H. 18	20	6	30.00	73	H. 56	16	5	31.25	73
H. 19	17	4	23.52	76	H. 57	13	8	61.53	59½
H. 20	23	8.5	35.41	70½					

TABLE XL.—*Ordination of 27 indices of alveolo-subnasal prognathism.—Salado.*

	No. of skull.	Index.		No. of skull.	Index.		No. of skull.	Index.
1	H. 43	14.28	10	H. 45	31.25	19	H. 14	44.44
2	H. 17	17.64	11	H. 56	31.25	20	H. 35	45.00
3	H. 3	21.05	12	H. 42	31.81	21	H. 7	46.42
4	H. 19	23.52	13	H. 10	32.50	22	H. 46	46.66
5	H. 25	27.27	14	H. 23	33.33	23	H. 2	50.00
6	H. 33	28.57	15	H. 20	35.41	24	H. 4	53.33
7	H. 18	30.00	16	H. 11	39.47	25	H. 37	56.25
8	H. 16	31.25	17	H. 54	41.17	26	H. 15	58.82
9	H. 24	31.25	18	H. 5	42.85	27	H. 57	61.53

Variation, 47.25. Theoretical mean of variation, 37.90. Skull nearest to mean, H. 11. Average, 37.27.

TABLE XLI.—*Ordination of 27 angles of alveolo-subnasal prognathism.—Salado.*

	No. of skull.	Angle.		No. of skull.	Angle.		No. of skull.	Angle.
		°			°			°
1	H. 57	59½	10	H. 5	67	19	H. 24	73
2	H. 15	60	11	H. 11	68	20	H. 45	73
3	H. 4	62	12	H. 54	68	21	H. 56	73
4	H. 37	62	13	H. 20	70½	22	H. 53	74½
5	H. 2	63½	14	H. 10	72	23	H. 25	75
6	H. 7	66	15	H. 23	72	24	H. 19	76
7	H. 14	66	16	H. 42	72½	25	H. 3	78
8	H. 35	66	17	H. 46	73	26	H. 17	79½
9	H. 46	66	18	H. 18	73	27	H. 43	82

Variation, 22½°. Theoretical mean of variation, 70¾°. Average, 70+°. Skull nearest to mean and average, H. 20

TABLE XLII.—*Seriation of 27 angles of alveolo-subnasal prognathism.—Salado.*

	Angle.	Number of skulls.		Angle.	Number of skulls.
1	59 to 60	1	13	71	0
2	60	1	14	72	3
3	61	0	15	73	5
4	62	2	16	74	1
5	63	1	17	75	1
6	64	0	18	76	1
7	65	0	19	77	0
8	66	4	20	78	1
9	67	1	21	79	1
10	68	2	22	80	0
11	69	0	23	81	0
12	70	1	24	82	1

Maximum of seriation, 73.

TABLE XLIII.—*Ordination of 38 orbital indices.—Salado.*

	No. of skull.	Index.		No. of skull.	Index.		No. of skull.	Index.
1	H. 22	81.81	14	H. 18	89.18	27	H. 41	94.66
2	H. 1	82.92	15	H. 8	89.61	28	H. 32	94.73
3	H. 6	84.61	16	H. 13	89.87	29	H. 37	94.73
4	H. 5	85.36	17	H. 25	90.24	30	H. 20	94.80
5	H. 19	85.71	18	H. 17	90.78	31	H. 11	94.87
6	H. 26	85.71	19	H. 51	90.90	32	H. 12	95.00
7	H. 21	85.89	20	H. 29	91.42	33	H. 27	95.00
8	H. 3	87.17	21	H. 43	91.66	34	H. 40	95.05
9	H. 57	87.80	22	H. 16	91.89	35	H. 4	97.29
10	H. 10	88.09	23	H. 56	92.10	36	H. 15	97.29
11	H. 15	88.15	24	H. 53	92.75	37	H. 14	98.64
12	H. 28	88.57	25	H. 24	93.42	38	H. 36	100.00
13	H. 7	88.75	26	H. 49	94.44			

Variation, 18.19. **Theoretical mean of variation, 90.90. Skull nearest to mean, H. 51.** Average, 91.10.

Among the above skulls, H. 19, H. 26, H. 21, H. 57, H. 7, H. 18, H. 25, H. 12, H. 40, H. 15, and H. 36, eleven in all, belong to **apparently normal skulls**; their average index is 91.06.

TABLE XLIV.—*Seriation of 38 orbital indices.—Salado.*

Index.	Number of skulls.	Index.	Number of skulls.
81 to 82	1	91	3
82	1	92	2
83	0	93	1
84	1	94	6
85	4	95	2
86	6	96	1
87	2	97	2
88	4	98	1
89	3	99	0
90	3	100 to 101	1

Maximum of frequency, 94.

TABLE XLV.—*Ordination of 44 nasal indices.—Salado.*

Number.	Index.		Number.	Index.		Number.	Index.
1	H. 43	44.23	16	H. 24	50.00	31	H. 29	53.57
2	H. 34	44.54	17	H. 2	51.06	32	H. 19	53.84
3	H. 14	45.28	18	H. 8	51.06	33	H. 18	54.00
4	H. 27	45.28	19	H. 32	51.06	34	H. 4	54.34
5	H. 52	45.28	20	H. 40	51.11	35	H. 37	54.73
6	H. 9	45.91	21	H. 42	51.92	36	H. 16	54.94
7	H. 6	46.00	22	H. 5	52.00	37	H. 28	55.55
8	H. 45	46.07	23	H. 25	52.08	38	H. 22	56.52
9	H. 54	47.05	24	H. 45	52.13	39	H. 23	56.66
10	H. 10	47.95	25	H. 3	52.94	40	H. 44	56.86
11	H. 7	48.00	26	H. 17	53.06	41	H. 36	59.34
12	H. 61	48.42	27	H. 26	53.12	42	H. 49	59.37
13	H. 21	48.92	28	H. 41	53.12	43	H. 20	60.00
14	H. 11	48.97	29	H. 56	53.26	44	H. 57	61.11
15	H. 25	49.09	30	H. 50	53.33			

Variation, 16.88. Theoretical mean of variation, 52.67. Skull nearest to mean, H. 3. Average, 51.66.

TABLE XLVI.—*Seriation of 44 nasal indices.—Salado.*

	Index.	Number of skulls.		Index.	Number of skulls.
1	44 to 45	2	10	53	7
2	45	4	11	54	4
3	46	2	12	55	1
4	47	2	13	56	3
5	48	4	14	57	....
6	49	1	15	58	....
7	50	1	16	59	2
8	51	5	17	60	1
9	52	4	18	61 to 62	1

Maximum of frequency, 53.
Skull, H. 30, with nose deflected to one side, as by a blow, has an index of 42.34.

TABLE XLVII.—*Anterior nasal spine.—Salado.*

No. of skull.	Descriptive No.	No. of skull.	Descriptive No.	No. of skull.	Descriptive No.
H. 1	2	H. 19	2	H. 36	1
H. 2	2	H. 20	2	H. 37	2
H. 3	3	H. 21	3	H. 42	1
H. 4	3	H. 22	2	H. 43	5
H. 5	2	H. 23	2	H. 44	4
H. 6	1	H. 24	1	H. 45	2
H. 7	3	H. 25	3	H. 46	3
H. 8	1	H. 26	1	H. 48	2
H. 10	3	H. 28	2	H. 49	2
H. 11	1	H. 29	2	H. 50	2
H. 12	2	H. 30	3	H. 54	1
H. 14	3	H. 32	3	H. 56	3
H. 16	2	H. 33	3	H. 57	2
H. 17	2	H. 34			
H. 18	1	H. 35	2		

TABLE XLVIII.—*Ordination of 32 palatine indices.—Salado.*

	No. of skull.	Index.		No. of skull.	Index.
1	H. 30	62. 74	17	H. 15	73. 07
2	H. 27	62.96	18	H. 56	73. 07
3	H. 25	63. 15	19	H. 17	74. 07
4	H. 20	64. 86	20	H. 8	74. 50
5	H. 1	65. 48	21	H. 33	75. 47
6	H. 5	66. 66	22	H. 49	76. 00
7	H. 37	67. 67	23	H. 40	76. 00
8	H. 35	69. 81	24	H. 50	76. 19
9	H. 10	70. 58	25	H. 57	76. 92
10	H. 11	70. 58	26	H. 46	77. 55
11	H. 43	71. 05	27	H. 29	77. 90
12	H. 24	71. 42	28	H. 28	78. 94
13	H. 45	71. 69	29	H. 41	80. 00
14	H. 14	71. 92	30	H. 4	81. 63
15	H. 2	73. 00	31	H. 16	81. 63
16	H. 7	73. 07	32	H. 19	84. 61

Variation, 21.87. Theoretical mean of variation, 73.67. Skull nearest to mean, H. 17. Average, 72.94.

If the aberrant skull H. 19 were excluded the variation would be 18.89; the mean, 72.18; the skull nearest to mean, H. 14, and the average, 72.57.

TABLE XLIX.—*Seriation of 32 palatine indices.—Salado.*

	Index.	Number of skulls.		Index.	Number of skulls.
1	62 to 63	2	13	74	2
2	63	1	14	75	1
3	64	1	15	76	4
4	65	1	16	77	2
5	66	1	17	78	1
6	67	1	18	79	0
7	68	0	19	80	1
8	69	1	20	81	2
9	70	2	21	82	0
10	71	4	22	83	0
11	72	0	23	84 to 85	1
12	73	4			

Maximum of frequency, 71, 73, and 76.

TABLE L.—*Ordination of palatine depth.—Salado.*

	No. of skull.	Palatine depth.		No. of skull.	Palatine depth.
1	H. 33	21	17	H. 57	15
2	H. 30	20	18	H. 4	14
3	H. 19	19	19	H. 11	14
4	H. 43	19	20	H. 16	14
5	H. 5	18	21	H. 27	14
6	H. 10	18	22	H. 45	14
7	H. 14	18	23	H. 1	13
8	H. 17 (immature)	18	24	H. 2	13
9	H. 20	17	25	H. 7	13
10	H. 24	17	26	H. 26	13
11	H. 41	17	27	H. 8	12
12	H. 31	16	28	H. 29 (child)	11
13	H. 37	16	29	H. 50 (child)	11
14	H. 40	16	30	H. 28 (child)	9
15	H. 49	16	31	H. 51 (child)	8
16	H. 15	15			

TABLE LI.—*Osteometrical measurements and indices of the long bones.—Salado.*

Designation of skeleton.	Humerus.		Radius.		Ulna.		Antibrachial index of right side.	Antibrachial index of left side.	Femur.		Tibia.		Fibula.		Tibio-femoral index of right side.	Tibio-femoral index of left side.
	Right	Left	Right	Left	Right	Left			Right	Left	Right	Left	Right	Left		
H. 1	281	278	216	215	283	297	76.59	77.39				338	c.321			*78.70
H. 5	282	280	239	237			79.13	79.00	426	431		254	347	341	*82.98	*82.13
H. 6	307	304	314	345	283	283	79.17	80.59	c.424	421	289		357	356	*84.66	*81.18
H. 7	312	311	246	247	264	c.265	78.84	79.45	431	427	309		326	360	*85.09	*85.07
H. 8	280	279	226	236	249	265	77.99	81.00	399	405	305		328	324	*85.03	*88.81

NOTE.—All the indices thus (*) marked were obtained by using in the calculation the average length of the respective series in place of the length of the missing bone.

TABLE LII.—*Synopsis of average indices of the long bones.—Salado.*

	Antibrachial.						Tibio-femoral.					
	Of right arm.		Of left arm.		Of total or both arms.		Of right leg.		Of left leg.		Of total of both legs.	
	Number of individual indices.	Average.	Number of individual indices.	Average.	Number of individual indices.	Average.	Number of individual indices.	Average.	Number of individual indices.	Average.	Number of individual indices.	Average.
Computed by method I	14	78.66	14	78.60	14	78.63	12	84.79	10	84.96	11	84.83
Computed by method II	18	77.81	12	79.17	15	78.35	11	85.87	13	92.45	12	84.02
Total	32	78.18	26	78.86	29	78.49	23	85.28	23	83.54	23	84.41

TABLE LIII.—*Antibrachial and tibio-femoral indices in various races.*

	Antibra-chial index.		Tibio-femoral index.
111 Europeans	72.47	9 Esthonians	78.60
9 Esthonians	73.10	6 Tartars	79.60
6 Egyptians	74.60	73 Europeans	81.62
6 Tartars	74.70	6 Chinese and Javanese	81.63
7 Arabs and Berbers	74.85	5 Polynesians	82.20
5 Kourgans in Russia	75.40	7 Arabs and Berbers	82.64
11 South Americans	75.77	42 African Negroes	83.26
5 New Caledonians	75.94	11 New Caledonians	83.48
7 Hindoos	76.25	11 South Americans	83.55
7 Chinese, Annamites, and Javanese	77.97	23 Saladoans	84.44
29 Saladoans	78.49		
42 African Negroes	78.83		

NOTE.—The number 29, referring to the Sala-doans, means 29 indices of various individuals, not the indices of 29 individuals.

NOTE.—The number 23, referring to the Sala-doans, means 23 indices of various individuals, not the indices of 23 individuals.

All the figures in the above table, except those concerning Saladoans, are from Topinard.[*] We have not copied all his figures, however, but only those which deal with five or more individuals. To make his data more comparable with ours we have combined the indices of the two sexes which he gives separately.

TABLE LIV.—*Dimensions and indices of 11 Scapulæ.—Salado*

Designa-tion of skeletons.	Length.		Width.		Indices.	
	Right scapula.	Left scapula.	Right scapula.	Left scapula.	Right scapula.	Left scapula.
H. 1	135cm.	....	93	....	68.84	....
H. 6	142	147	99	101	69.71	68.70
H. 8	136	137	97	101	71.32	73.72
H. 21	....	182	....	102	....	67.10
H. 25	161	160	105	110	65.21	68.75
H. 33	130	....	94	....	72.30	....
H. 36	130	....	98	....	81.66	....
H. 45	....	138	....	105cm.	....	74.63
H. 68	132	130	90	92	68.18	70.76
H. 70	123	....	97	....	78.86	....
H. 72	162	....	108	....	66.66	....
Average index					71.42	70.61

General average index, 71.09.

TABLE LV.—*Angle of torsion of humerus.—Ordination according to right humerus.—Salado*

No.	Sex.	Right.		Left.		No.	Sex.	Right.		Left.	
		°	′	°	′			°	′	°	′
1	....	146	00	148	30	12	....	160	00	....	
2	....	148	00	155	00	13	H. 15	F.	160	30	154	00
3	H. 26	M.	148	30	....		14		M.	161	00	....	
4	H. 5	F.	151	30	154	00	15	H. 41(?)	M.	161	30	153	00
5			153	30	166	00	16	H. 8	F.	161	30	162	30
6		F.	155	00	157	30	17	H. 25	M.	166	00	166	00
7	H. 33		155	00	....		18	H. 6	M.	170	00	168	30
8	H. 39	F.	157	00	169	00	19	H. 33		174	00	....	
9	H. 19	M.	158	30	156	00	20	H. 45		174	00	169	00
10	H. 1	F.	158	30	152	00	21		....	177	00	....	
11	H. 32	M.	158	30	161	30						

Average of 21 right humeri = 159° 15′.
Average of 41 humeri, both sides = 159° 30′ 4.

[*] TOPINARD: *Op. cit.*, pp. 1043, 1045.

TABLE LVI.—*Angle of torsion of humerus.—Ordination according to left humerus.—Salado.*

Number.	Sex.	Left.	Right.	Number.	Sex.	Left.	Right.
		° '	° '			° '	° '
1		148 30	146 00	11	H. 32	M.	161 30	158 30
2	H. 19	M.	150 00	158 30	12		161 30	
3	H. 5	F.	151 00	151 30	13	H. 8	F.	162 30	161 30
4	H. 1	F.	152 00	158 30	14		164 00	
5	H. 7	M.	152 30		15		166 00	153 30
6	H. 41(?)		153 00	161 30	16	H. 25	M.	166 00	196 00
7	H. 15	F.	154 00	160 30	17	H. 6	M.	168 30	170 00
8		155 00	148 00	18	H. 39	F.	169 00	157 00
9	H. 57		157 00		19	H. 45		169 00	174 00
10		F.	157 30	155 00	20		172 00	

Average of 20 left humeri = 159° 30'.

TABLE LVII.—*Mean angles of torsion of pairs of humeri.—Salado.*

	Designation.	Sex.	Mean angle of pair.	Variation between right and left humerus.
			° '	° '
1			147 15	2 30
2	H. 5	F.	151 15	30
3			151 30	7 00
4	H. 19	M.	154 15	8 30
5	H. 1	F.	155 15	6 30
6		F.	156 15	1 30
7	H. 41(?)		157 15	8 30
8	H. 15	F.	157 15	6 30
9			159 45	12 30
10	H. 32	M.	160 00	3 00
11	H. 8	F.	162 00	1 00
12	H. 39	F.	163 00	12 00
13	H. 25	M.	166 00	
14	H. 6	M.	169 15	1 30
15	H. 45		171 30	5 00

Average mean angle of 15 pairs = 158° 47'.
Average variation between right and left humerus, 5° 6'.

TABLE LVIII.—*Torsion of the humerus with regard to sex.—Salado.*

Male.			Female.		
Designation of skeleton.	Right humeri.	Left humeri.	Designation of skeleton.	Right humeri.	Left humeri.
	° '	° '		° '	° '
H. 19	158 30	150 00	H. 5	151 30	151 00
H. 32	158 30	161 30		155 00	157 30
H. 25	166 00	166 00	H. 39	157 00	169 00
H. 6	170 00	168 30	H. 1	158 30	152 00
			H. 15	160 30	154 00
			H. 8	161 30	162 30
Average...	163 15	161 30	Average...	157 20	157 30

Total average of males, 162° 22'.
Total average of females, 157° 30'.

TABLE LIX.—*Torsion of the humerus in specimens in the general collection of the* **Army Medical Museum**.

	Museum No.	Race or nation.	Sex.	Left.	Right.
				°	°
1	165	Mound Builder	M.	150.00	152.50
2	2621	.....do	....	148.00	152.00
3	1530	Apache	M.	154.25	135.50
4	1790	Sioux, Sisseton	F.	158.50	154.25
5	1791	.....do	M.	162.75	155.25
6	1792	.....do	M.	156.75	154.25
7	1832	Sioux, Brulé	M.	144.00	141.00
8	1896	.....do	M.	156.00	148.75
9	1897	.....do	F.	155.00	156.75
10	1898	.....do	M.	160.25	154.00
11	1901	.....do	M.	150.50	136.75
12	2062	.....do	M.	154.75	154.50
13	2968	.....do	F.	151.00	152.75
14	1855	Sioux, Ogalalla	F.	142.25	151.50
15	1854	.....do	F.	150.50	144.60
16	2046	Sioux	F.	156.50	156.75
17	2071	.....do	M.	157.00	157.50
18	2072	.....do	F.	157.00	158.50
19	2836	Cheyenne	M.	153.25	150.25
20	2966	.....do	M.	154.50	142.75
21	6409	Arapahoe	M.	168.25	162.75
22	1305	Peruvian	F.	161.75	162.00
23	1	French	M.	174.00	179.00
24	121	.....do	F.	168.25	166.75
25	1578	.....do	M.	170.00	159.50
26	1581	.....do	M.	178.50	175.25
27	1585	.....do	F.	175.50	172.50
28	1620	.....do	M.	176.00	175.25
29	2135	.....do	F.	179.25	160.00
30	2372	Lapp	F.	161.75	160.25
31	2373	.....do	....	173.75	164.25
32	2374	.....do	....	167.00	161.50
33	2375	.....do	....	165.50	161.75
34	2376	.....do	....	165.75	166.50
35	2377	Finn	....	158.50	153.00
36	959	Chinese	M.	159.50	156.00
37	1835	Malatto	M.	157.00	151.00
38	2040	.....do	F.	161.00	165.00
39	552	Negro	F.	157.00	140.50
40	1021	.....do	F.	150.25	153.00
41	2837	.....do	F.	160.00	158.50
42	2941	.....do	M.	169.75	161.25
43	2103	.....do	M.	148.50	150.50
		Average angle of American Indians		154.67	151.96
		Average angle of all others		165.20	161.98
		Average angle of total		159.81	156.63

TABLE LX.—*Torsion of the humerus in specimens in the general collection of the* **Army Medical Museum** *(series less than five excluded).—Average for male and female and for right and left.*

Race or nation.	Number of individuals.	Female.			Male.			Undetermined sex.			Right total.	Left total.	Total of all.
		Right.	Left.	Total.	Right.	Left.	Total.	Right.	Left.	Total.			
French	3 females. 4 males.	168.41	171.85	169.12	172.50	174.62	173.50				169.89	173.42	171.66
Lapps	1 female. 4 undetermined.	160.25	161.75	161.00				163.00	168.00	165.75	162.85	166.75	164.85
Sioux or Dakotas	7 females. 8 males.	150.50	155.06	153.25	150.25	155.25	152.75				151.70	154.18	152.97
Other North Americans	5 males. 1 undetermined.				150.55	156.05	153.30	152.00	148.00	150.05	150.79	154.79	152.79
Negroes	5 females. 2 males.	156.00	156.75	153.20	155.87	159.12	157.50				152.70	157.12	154.92

TABLE LXI.—*Indices of 19 pelves—Salado.*

Designation of skeleton.	Breadth-height index.	Superior strait index.	Pubo-ischiatic depth index.	Sacral length index.
H. 1	....	77.95	65.35	72.44
H. 5	135.32	84.72	66.06	77.08
H. 6	136.36	91.45	76.06	94.87
H. 7	145.83	74.04	71.75	67.93
H. 8	142.32	85.82	66.14	80.31
H. 10	148.92	80.41	66.43	65.03
H. 14	131.77	103.44	77.58	83.62
H. 15	152.09	74.61	59.23	73.84
H. 18	....	86.03	69.99	....
H. 19	....	89.65	81.89	95.08
H. 25	140.60	82.81	81.25	....
H. 36	149.42	76.92	70.76	76.92
H. 39	152.38	69.04	66.60	....
H. 41	131.18	78.63	80.31	88.52
H. 45	....	82.26	63.12	73.75
H. 57	152.66	82.53	64.28	76.98
H. 59	137.07	69.06	62.58	66.18
H. 72	146.96	80.14	69.85	82.33
H. 96	130.17	78.21	75.24	89.10

TABLE LXII.—*Ordination of breadth-height indices of 14 pelves—Salado.*

	Designation of skeleton.	Index.	Sex.		Designation of skeleton.	Index.	Sex.
1	H. 41	131.18	Male.	8	H. 7	145.82	Male.
2	H. 14	131.77	Male.	9	H. 72	146.96	Male.
3	H. 5	135.32	Female.	10	H. 10	148.92	Female.
4	H. 6	136.36	Male.	11	H. 36	149.42	Female.
5	H. 59	137.07	Female.	12	H. 15	152.09	Female.
6	H. 25	140.60	Male.	13	H. 39	152.38	Female.
7	H. 8	142.32	Female.	14	H. 57	152.66	Female.

TABLE LXIII.—*Ordination of superior strait indices of 18 pelves.—Salado*

	Designation of skeleton.	Index.	Sex.		Designation of skeleton.	Index.	Sex.
1	H. 39	69.04	Female.	10	H. 45	82.26	Female.
2	H. 59	69.06	Female.	11	H. 57	82.53	Female.
3	H. 7	74.04	Male.	12	H. 25	82.81	Male.
4	H. 15	74.61	Female.	13	H. 5	84.72	Female.
5	H. 36	76.92	Female.	14	H. 8	85.82	Female.
6	H. 1	77.95	Female.	15	H. 18	86.03	Male.
7	H. 41	78.63	Male.	16	H. 19	89.65	Male.
8	H. 72	80.14	Male.	17	H. 6	91.45	Male.
9	H. 10	80.41	Female.	18	H. 14	103.44	Male.

TABLE LXIV.—*Ordination of 18 pubo-ischiatic indices.—Salado.*

Designa-tion of skel-eton.	Index.	Sex.	Designa-tion of skel-eton.	Index.	Sex.
1	H. 15	59, 23	Female.	10	H. 18	68, 99	Male.
2	H. 59	62, 58	Female.	11	H. 72	69, 85	Male.
3	H. 45	63, 12	Female.	12	H. 36	70, 76	Female.
4	H. 57	64, 28	Female.	13	H. 7	71, 75	Male.
5	H. 1	65, 35	Female.	14	H. 6	76, 06	Male.
6	H. 8	66, 14	Female.	15	H. 14	77, 58	Male.
7	H. 10	66, 43	Female.	16	H. 41	80, 34	Male.
8	H. 5	66, 66	Female.	17	H. 25	81, 25	Male.
9	H. 39	66, 66	Female.	18	H. 19	81, 89	Male.

TABLE LXV.—*Ordination of 15 sacral length indices.—Salado.*

Designa-tion of skel-eton.	Index.	Sex.	Designa-tion of skel-eton.	Index.	Sex.
1	H. 10	65, 03	Female.	9	H. 5	77, 68	Female.
2	H. 59	66, 18	Female.	10	H. 8	80, 31	Female.
3	H. 7	67, 93	Male.	11	H. 72	82, 35	Male.
4	H. 1	72, 44	Female.	12	H. 14	83, 64	Male.
5	H. 45	73, 75	Female.	13	H. 41	86, 32	Male.
6	H. 15	73, 84	Female.	14	H. 6	94, 87	Male.
7	H. 36	76, 92	Female.	15	H. 19	95, 68	Male.
8	H. 37	76, 98	Female.				

TABLE LXVI.—*Breadth height indices (general index of the pelvis) in various races.*

Males.		Females.	
6 Saladoans	138. 78	8 Saladoans	146, 27
46 Europeans	126. 66	24 Europeans	136, 90
17 African negroes	121, 30	10 African negroes	134, 20
11 Oceanian negroes	122, 70	10 Oceanian negroes	129, 00

NOTE.—With the exception of the Saladoans these data are from TOPINARD's Éléments d'anthropologie, p. 1049.

TABLE LXVII.—*Indices of the superior strait in various races.*

Males.		Females.	
63 Europeans	80, 00	10 Saladoans	78. 33
2 Lapps	83, 00	49 Europeans	79, 00
8 Saladoans	85, 77	6 African negresses	81, 00
1 Tasmanian	88, 00	3 Peruvians	83, 00
17 African negroes	89, 00	7 Australians	86, 00
12 New Caledonians	91, 00	3 New Caledonians	89, 00
1 Australian	98, 00	1 Javanese	90, 00
		13 Andamanese	99, 00

NOTE.—With the exception of the figures on the Saladoans, these data are from TOPINARD's Éléments d'anthropologie, p. 1050.

TABLE LXVIII.—*19 different measurements of 30 pelves.—Salado.*

Designation of skeletons.	Conjugata externa.	Crest width.	Antero-superior spinal width.	Posterior-superior spinal width.	Antero-posterior diameter of brim.	Transverse diameter of brim.	Antero-posterior diameter of outlet.	Transverse diameter of outlet.	Sciatic width.	Pelvic height. Right.	Pelvic height. Left.
H. 1......	161				99	127	124	100	91 ca.	170	
H. 5......	188	272	233	163	122	144 ca.	118	120	111	201	202
H. 6......	163	255	220	72	107	117	101 ca.	87	78	187	190
H. 7......	155	280	249		97 ca.	131	118	93	96	192	193
H. 8......	161	289	234	77	109	127	113	110	99	189	181
H. 10......	187 ca.	277	244	92 ca.	115	143	150	112	102	186	188
H. 14......	174 ca.	253	205		120	116	108	92	82 ca.	192	188
H. 15......	154	245	211	77	97	130	110	123		167	168
H. 18......	165 ca.		224		111 ca.	129		100 ca.	87 ca.	192	
H. 19......					104	116	88	77			202
H. 21......										209	206
H. 25......	158 ca.	277	241 ca.		106	128					197
H. 27......											
H. 33......										176	
H. 35......											
H. 36......	198	260	229	88	100	130	108 ca.	107		174	173
H. 39......		256	245		87	126		103		168	
H. 41......	149	244 ca.		86	92	117	97	90		186	186
H. 45......	165			92	118	111	117	125	114		195
H. 57......	159	258	234		104	126	114	106	101		169
H. 59......	169	244	227	85	96	139	119	109	108	178	179
H. 61......											
H. 63......											
H. 72......	154 ca.	291	263		109	136	105	107	99 ca.		198
H. 79......											
H. 84......											
H. 87......										187	
H. 96 (child, epiphyses absent)	130	220	204	72 c.	79	101	94	69		189	169
B......											
C......											

Designation of skeletons.	Iliac breadth. Right.	Iliac breadth. Left.	Height of iliac fossa. Right.	Height of iliac fossa. Left.	Cord of the brim.	Pubo-ischiatic depth. Right.	Pubo-ischiatic depth. Left.	Acetabulo-symphyseal width.	Sacral length.	Sacral breadth.	Width of sacrum at brim.	Inferior width of sacrum.
H. 1......					106		83 ca.	113	92	101	92	83
H. 5......		152	89		131	96	96	128	111	126	107	101
H. 6......	146		89	90	107 ca.	89	91	110	111	116	100	83
H. 7......	153		89	89	106	94	95	113	89	117	110	91
H. 8......	141	144	83	83	108	84	84	113	102	122	105	81
H. 10......	154		88	90	127	95	95	131	93 (5 v)	126	100	96
H. 14......			92	87	118	90	90	119	97	120	98	
H. 15......	139	146	85	90	105	77 ca.	76	112	96	103	101	83
H. 18......		121 ca.	82		112	89		120		110	100	87
H. 19......					108		95	115	111	122	99	90
H. 21......			98 ca.	96 ca.	115	98	100	122				
H. 25......	160 ca.			96	112 ca.		104	129			104	
H. 27......									112	118	100	92
H. 33......			82		113	88		108	99 (5 v)	111	99	86
H. 35......									103	113	109	86
H. 36......	133	131 ca.	82	79	114	92	90	120	100 (5 v)	117	105	86
H. 39......	133		82 ca.		105	84	84	112		106	101	
H. 41......	132		88	89 ca.	100	94	89 ca.	105	101	108	99	74
H. 45......	152 ca.	149		95	106		89	119	104	122	109	95
H. 57......			85	83	105	81	82	112	97	113	104	82
H. 59......	154	136	81	83	121	87	87	114	92	111	104	91
H. 61......									105			93
H. 63......									102 (5 v)	117 ca.	110	92 ca.
H. 72......		163		99	111	95		120	112	129	115	96
H. 79......									108 ca.	106	94	80 ca.
H. 84......									96	121	115	90
H. 87......	144		87						105	117 ca.	98	88
H. 96 (child, epiphyses absent)	115	116	72	74	91	76	76	89 ca.	90	92	88	70
B......									97	121	99	89
C......									92	111	104	90

TABLE LXIX.—*Pilaster femur—Indices of transverse section of shaft of femur from 54 skeletons, more or less complete.—Salado.*

Designation of skeletons	Right, 49.			Left, 47.			Remarks.
	Antero-posterior diameter.	Lateral diameter.	Index.	Antero-posterior diameter.	Lateral diameter.	Index.	
H. 1	23½	23	102.17	23½	23½	100.00	Both diameters a little oblique.
H. 2	30	23	130.43				
H. 5	30½	23	132.60	30	24	125.00	
H. 6	30	21½	139.53	31	21	147.61	
H. 7	32	24	133.33	30	23½	127.65	
H. 8	27½	23	119.56	29	24	120.83	
H. 9	24	24½	97.95	23½	24	97.91	
H. 10	27½	26½	103.77	28	25½	109.80	
H. 15	24	21½	111.62	23½	21½	109.30	
H. 19	30	24½	122.44	31	25	124.00	Slight exostosis in left.
H. 21	32	27½	116.36	33	27	122.22	
H. 25	32	23	138.13	31½	24	131.25	
H. 29	17	13	130.76	16½	13	126.92	Young child.
H. 30	29	25	116.00	31	23	134.78	Femora bowed forward.
H. 32	28	24	116.66	27½	23	119.56	
H. 33				24	21	114.28	
H. 34	30	23½	127.65	30	25	120.00	
H. 36	27	23	117.39	26½	23	115.21	
H. 39	23	22	104.54	23	21	109.52	
H. 41	25½	22½	113.33	25½	24	104.25	
H. 42	33	24½	134.69	32	25	128.00	Exostotic fringes on linea aspera.
H. 45	26	22½	115.55	26½	22½	117.77	
H. 57	25	22	113.63	25	22	113.63	
H. 58				29	25½	113.72	
H. 59	24	21½	111.62	24	20½	117.07	
H. 60	25	22	113.63	25	21	119.04	
H. 62	30	27	111.11				
H. 63	31	25	124.00	31	24½	126.53	Slight exostosis on right.
H. 64	27	29	93.10				
H. 65	29	25	116.00				
H. 66	24½	23½	104.25				
H. 67	25	25	100.00				
H. 69	16½	14	117.85	16	14	114.28	Child.
H. 70	24½	23	106.52	25½	23½	108.51	
H. 71				27	21	128.57	
H. 72	30½	23½	129.78	30	24	125.00	
H. 73	30½	24	127.08	29	23½	123.40	
H. 74	24	24½	97.95	24½	24½	100.00	
H. 75	26	24	108.33	26	25	104.00	
H. 76	24	22	109.09	24	22	109.09	
H. 77	25½	23	110.86	26½	23	115.21	
H. 78	27	24½	110.20				
H. 79	28½	26½	107.54	29	25½	113.72	
H. 81				25½	24	106.25	
H. 82	24	23	104.34	25	22½	111.11	
H. 85	25½	23½	108.51	26½	25	106.00	
H. 86	33	24½	131.63	30½	26½	115.09	
H. 87	30½	23	132.60	30	23	130.43	
H. 88				29½	23	128.26	
H. 90	28	25	112.00	30	23½	127.65	
H. 91	29	27	107.40	29	27	107.40	
H. 92	28	23	121.73	27	24½	110.20	
H. 93	26	22½	115.55	25½	23	110.86	
H. 96	19	19½	97.43	23	18½	124.32	Youth.
Average index			115.76			117.38	

Average index of total, 116.45.

TABLE LXX.—*Pilaster femur—Indices of transverse section of shaft of femur, miscellaneous,—Salado.*

Designation of skeleton	Right, 17.			Left, 16.			Remarks.
	Antero-posterior dimension.	Lateral dimension.	Index.	Antero-posterior dimension.	Lateral dimension.	Index.	
S 1	27	24	112.50	28	23	121.73	
S 11	25½	25½	100.00	25	24½	102.04	
A	29½	20½	143.90	28	21	133.33	
B	25	22½	111.11	24	23	104.34	
C	27	27	100.00	27	26	103.84	
D	17½	16½	106.06	18½	16	115.62	Child.
E				28	22	127.27	
F	26	23	113.04	25½	23	110.86	
G	25	23	108.69	26	22½	115.55	
H	26	23	113.04				
I	31	27	114.81	31	28½	108.77	
K	26	24½	106.12				
L				33	25½	129.41	
M	25½	22	115.90				
N	27	25	108.00				
O				30½	25½	115.68	
P				27	24	112.50	
Q				25	20	120.00	
R	26½	23½	112.76				
S				25	23	108.69	
T				20½	22½	111.11	
U				24	20	120.60	Both diameters a little oblique.
V	32	25½	125.45				
W	25	24	104.16				
X				21	17	123.52	Child.
Y	20½	19½	105.12				
Average index			111.80			115.79	

Average index of total, 113.85.

TABLE LXXI.—*Indices of section of the femur in 16 Peruvian skeletons in Army Medical Museum.*

Number of specimen	Right.			Left.			Remarks.
	Antero-posterior dimension.	Lateral dimension.	Index.	Antero-posterior dimension.	Lateral dimension.	Index.	
1595	23	22½	102.22	23½	24	97.91	Female.
2555	30	27½	110.90				
3131	22	20	110.00	22½	22	102.27	Adolescent.
3132	27	29½	91.52	25½	29½	86.44	
446	24	22½	106.66	24½	23	106.52	
447	31½	24½	128.57	32	25	128.00	
448	25½	26	98.07	29	27	107.40	
449	27	24½	110.20	26½	23½	112.76	
450	30	29	103.44	32	27	118.51	
451	27½	25	110.00	27½	25	110.00	
452	27½	28½	96.49	28	28½	98.23	
453	24	22	109.09	23½	23½	100.00	
454	25	25½	98.03	25½	25½	100.00	
455	31	26	119.23	30½	26	117.30	
456				31½	29	108.62	
457				33	30	110.00	
Average index			106.74			106.93	

Specimens of provisional section, A.M.M. bracketing rows 446–457.

Average index of total, 106.84.

TABLE LXXII.—*Of indices of transverse section of shaft of femur—entire collection of Salado.*

Average of 66 femora of right side................................... 114. 74
Average of 65 femora of left side 116. 94
Average of 131 femora of both sides 115. 83
Maximum (H. 6, left) 147. 64
Minimum (H. 64, right) 93. 10

TABLE LXXIII.—*Relations existing between pilaster femur and platycnemic tibia.—Salado.*

[The first 5 skeletons have the lowest average tibial index and the highest average femoral index. In the last 5 the conditions are reversed.]

Designation of skeleton.	Tibiæ.			Femora.		
	Right.	Left.	Both.	Right.	Left.	Both.
Lowest						
H. 6	51. 42	49. 29	50. 35	139. 53	147. 64	143. 57
H. 19	48. 75	48. 75	48. 75	122. 44	124. 00	123. 22
H. 21	59. 15	50. 60	54. 87	116. 26	122. 22	119. 29
H. 30	54. 34	51. 47	53. 00	116. 00	134. 78	125. 39
H. 87	52. 11	50. 00	51. 05	132. 60	130. 43	131. 51
Total average	53. 19	50. 02	51. 60	125. 38	131. 80	128. 59
Highest						
H. 5	72. 33	70. 49	71. 91	132. 60	125. 00	128. 80
H. 36	76. 28	75. 43	75. 85	117. 39	115. 21	116. 30
H. 65	62. 16	68. 57	65. 36	124. 00	126. 53	125. 26
H. 70	69. 49	68. 75	69. 12	106. 52	108. 51	107. 51
H. 74	74. 19	79. 05	76. 61	97. 95	100. 00	98. 97
Total average	71. 09	72. 45	71. 77	115. 49	115. 65	115. 37

TABLE LXXIV.—*Platycnemia.—Indices of transverse section of shaft in 90 tibiæ from 52 skeletons more or less complete.—Salado.*

Designation of skeleton.	Right side.			Left side.			Remarks.
	Antero-posterior dimension.	Lateral dimension.	Index.	Antero-posterior dimension.	Lateral dimension.	Index.	
H. 1	26½	16	60. 37	27	17	62. 96	
H. 2	....	....	....	33	21	63. 63	
H. 5	30	22	73. 33	30½	21½	70. 49	
H. 6	35	18	51. 42	35½	17½	49. 29	
H. 7	36	19	52. 77	35	20	57. 14	
H. 9	29	18½	63. 79	30½	20½	67. 21	
H. 10	34	20	58. 82	34	22	64. 70	
H. 11	30	20	66. 66	32	20	62. 50	
H. 15	27	17	62. 96	26½	16½	62. 26	
H. 19	40	19½	48. 75	40	19½	48. 75	
H. 21	35½	21	59. 15	41½	21	50. 60	Muscular exostosis.
H. 26	25	19	76. 00	28	20	71. 42	Youth.
H. 29	20	13½	67. 50	18½	14	75. 67	Child.
H. 30	33	18	54. 54	34	17½	51. 47	
H. 32	....	....	....	31	21½	69. 35	
H. 33	27	17	62. 96	....	....	....	
H. 34	30½	23	64. 78	34½	23	66. 66	
H. 36	28	21½	76. 28	28½	21½	75. 43	
H. 39	27	15	55. 55	26½	15	56. 60	
H. 41	32½	21	64. 61	32	20	62. 50	
H. 42	37	22½	60. 81	35½	22	61. 97	
H. 45	31½	19½	61. 90	31	21	67. 74	
H. 48	36	24½	68. 72	....	....	....	
H. 57	27	17	63. 96	29	18	62. 06	
H. 58	32½	23	70. 76	33½	24½	73. 13	
H. 59	....	....	....	26½	18	67. 92	
H. 60	....	....	....	31	19½	62. 90	
H. 61	38	22	57. 89	....	....	....	
H. 62	36½	21	57. 53	....	....	....	
H. 63	37	23	62. 16	35	24	68. 57	
H. 64	36½	19½	53. 42	37	20½	55. 40	
H. 65	37	21½	58. 10	36	20½	56. 94	
H. 67	....	....	....	29	19	65. 51	

TABLE LXXIV—*Continued.*

Designation of skeleton.	Right side.			Left side.			Remarks.
	Antero-posterior dimension.	Lateral dimension.	Index.	Antero-posterior dimension.	Lateral dimension.	Index.	
H. 68	....	....	....	27½	19	69.09	
H. 69	21	15	71.42	19	14½	76.31	Child.
H. 70	29½	20½	69.49	32	22	68.75	
H. 71	30½	20½	67.21	....	....	....	
H. 72	....	....	....	35½	22	61.97	
H. 73	38½	21	54.54	36	20	55.55	
H. 74	31	23	74.19	31	24½	79.03	
H. 77	29	20	68.96	....	....	....	
H. 78	31	20	64.51	33	19	57.57	
H. 79	34	20	58.82	32½	21	64.61	
H. 81	31½	17	53.96	32	17	53.12	
H. 82	25½	17	66.66	25	18½	74.00	Slight exostosis.
H. 84	32½	21	64.61	31	19	61.29	
H. 85	....	....	....	29	20	68.96	
H. 86	38½	20	51.94	37½	19½	52.00	
H. 87	35½	18½	52.11	36	18	50.00	
H. 88	....	....	....	39	22	56.41	
H. 89	29	16½	56.89	....	....	....	
H. 90	33	17½	53.03	....	....	....	Deformed.
....	34½	24	69.56	37	25½	68.91	
....	....	....	....	15½	12	77.41	Child.
Average	....	....	61.78	....	....	63.60	

Average of all tibiæ, 62.71. Average of 78 normal adult tibiæ, 61.88.

TABLE LXXV.—*Platycnemia.—Indices of transverse section of shaft in 26 miscellaneous tibiæ.—Salado.*

Designation of skeleton.	Right side.			Left side.			Remarks.
	Antero-posterior dimension.	Lateral dimension.	Index.	Antero-posterior dimension.	Lateral dimension.	Index.	
A	....	....	....	39	26	66.66	
B	....	....	....	27½	19	69.09	
C	....	....	....	34	19	55.88	
D	....	....	....	32	22	68.75	
E	....	....	....	28½	19½	68.42	
F	29	17½	60.34	....	....	....	
G	37½	21	56.00	....	....	....	
H	35	22	62.85	....	....	....	
I	38	21	55.26	....	....	....	
K	....	....	....	29½	21	71.18	
L	....	....	....	30½	17	55.73	
M	36½	22½	61.64	....	....	....	
N and O	40	23	57.50	39	24½	60.28	Pair.
P	....	....	....	27	19	70.37	
Q and R	22½	15	66.66	22	15	68.18	Child.
S and T	21	17	80.95	21	17	80.95	Child.
U	14	12½	89.28	....	....	....	Very young child.
V and W	29	19	65.51	29	19	65.51	Probably a pair.
X	....	....	....	30	19½	65.00	
Y	28	20	71.42	....	....	....	
Z and A¹	28½	20½	71.92	29	20	68.96	Probably a pair.
Average	....	....	66.61	....	....	66.78	

Average of total, 66.70. Average of 116 tibiæ of both series (regular and miscellaneous), 63.54.

TABLE LXXVI.—*Indices of transverse section of shaft of femur in 62 skeletons of various races in the Army Medical Museum.*

Races.	No. of specimens.	Sex.	Right femur.			Left femur.			Remarks.
			Antero-posterior dimension.	Lateral dimension.	Index.	Antero-posterior dimension.	Lateral dimension.	Index.	
White	5433	M.	24	25½	94.11	23½	25½	92.15	Hunchback.
Do	6914	F.	27½	24	114.58	27½	24½	112.24	
Negro	552	F.	23	19½	117.94	24	20½	117.07	15 years of age.
Do	2057	F.	31	25½	121.56	31	25½	121.56	70 years of age.
Do	2041	M.	29½	28	105.35	29½	28	105.35	
Do	2103	M.	28½	27	105.55	29	27	107.40	
Do	3301	F.	25	22	113.63	26	22	118.18	Rarely an adult.
Do	5432	F.	38½	24½	159.18	37	25½	145.09	Hunchback.
Mulatto	1835	M.	32½	27½	118.18	33	28½	115.78	
Do	2046	M.	27	23½	114.89	27	23½	114.89	
Mexican	1834	M.	25	21½	116.27	24½	22	111.36	Youth.
Mahlemut Eskimo	756					25	26	96.15	Tumor on right femur.
Patagonian	239	F.	26	23½	110.63	25	22½	111.11	
North American Indians.									
Alaskan	814	M.	31½	25½	123.52	30½	26	117.30	
Do	938	M.	31	27½	112.72	31	26½	116.98	Hunchback.
Apache	1473	F.	22½	22½	100.00	22½	23	97.82	
Do	1530	M.	29	25½	113.72	30½	27	112.96	
Arapaho	6499	M.	29	27½	105.45	29½	24½	120.40	
Bannock	2133	M.	31½	26½	118.86	31½	25½	123.52	
Cheyenne	2036	M.	30	27	111.11	30½	27½	110.90	
Do	2066	M.	32½	27½	118.18	31½	27½	114.54	
Do	6866	M.	30	26½	113.20	31½	26½	118.86	Adolescent.
Chippewa	1287	M.	28½	26	109.61	29½	26½	111.32	
Choctaw	627	F.	29	24	120.83	28½	24½	116.32	
Do	629	F.	29½	23½	112.76	29	24	120.83	
Comanche	1000	F.	24	23	104.34				Exostosis of left femur.
Dakota	913	M.	32½	28½	114.63	31	29½	111.86	
Modoc	6287	M.	29½	26	113.46	29	25½	113.72	
Mound-builders:									
From Dakota	165	M.	32	29	110.34	29½	28½	105.59	
Do	1121	M.	32½	27	120.37				
From Mississippi	400	M.	30	32	93.75	28½	33	86.36	
Do	644	M.	29	23	126.08	28½	24	118.75	
Navajo	708	M.	26½	23	115.21	26	24½	106.12	
Pawnee	778	F.	27	23½	114.89	28	25½	109.80	
Pah Utes	963	M.	30½	24½	124.48	30½	26	122.00	
Do	964	F.	24	23½	102.12	23	24	95.83	
Sioux	13	M.	31	28½	108.78	32	26	123.07	
Do	2016	F.	26½	27	98.14	27½	28½	96.49	
Do	2071	M.	33	26½	124.52	32½	26	125.00	
Do	2072	F.	32	26½	120.75	33½	29	115.51	
Brule Sioux	1852	M.	29	26½	109.43	28½	25½	111.76	
Do	1895	M.	31	25½	121.56	31½	28½	118.26	
Do	1896	M.	31½	28½	121.05	35	30½	114.75	
Do	1897	F.	32	26	123.07	32	28	114.30	
Do	1898	M.	32	28½	112.28	32	29	110.34	
Do	1899	M.	36½	28	130.35	36	30½	118.03	
Do	1900	F.	25½	24	106.25	25½	25½	100.00	30 years of age.
Do	1901	M.	33	30½	108.19	33	29	113.79	
Do	2061	M.	31	29	106.89	30	28½	105.26	Adolescent.
Do	2082	M.	31½	27½	125.45	33½	29	115.51	
Do	2083	M.	24½	22	111.36	24½	23	106.52	Adolescent.
Do	2098	F.	28½	27½	105.63	28½	27	105.55	
Ogalalla Sioux	1851	F.?	23	21	109.52	23	22½	102.22	About 13 years of age.
Do	1853	F.	30½	26½	115.00	31	28	110.71	
Do	1854	F.	27	27	100.00	27½	29½	93.22	
Do	1856	M.	21½	21	102.38	21	20½	102.43	About 16 years of age.
Sisseton Sioux	1790	F.	27	27½	98.18	26	26	100.00	
Do	1791	F.	31½	27½	114.54	31	26	119.23	
Do	1792	M.	32	30½	104.91	33	31	106.45	
Yankton Sioux	926	M.	34½	28	123.21	34½	29	118.96	
Asiatics.									
Chinese	956	M.	29½	31½	93.65	30	30	100.00	
Do	957	F.	24	21½	111.62	24	23	104.34	

TABLE LXXVII.—*Platycnemia.—Indices of transverse section of shaft of tibia* **in 62 skeletons of** *various races in the Army Medical Museum.*

Races.	No. of specimen.	Sex.	Right tibia. Antero-posterior dimension.	Right tibia. Lateral dimension.	Right tibia. Index.	Left tibia. Antero-posterior dimension.	Left tibia. Lateral dimension.	Left tibia. Index.	Remarks.
White	5433	M.	28½	23	80.70	29	23	79.31	Hunchback.
Do	6414	F.	30½	23	75.40	28	23	82.14	
Negro	552	F.	27	19	70.37	28	19½	69.64	15 years of age.
Do	2037	F.	30½	22½	73.77	33	22½	68.18	70 years of age.
Do	2041	M.	34	26½	77.94	35	27	77.14	
Do	2103	M.	35	27½	78.57	35½	27½	77.46	
Do	3301	F.	30	22	73.33	31	22	70.96	Barely an adult.
Do	5432	F.	35½	27	76.05	35½	26½	74.64	Hunchback.
Mulatto	1835	M.	35	25	71.42	34½	24	69.56	
Do	2040	M.	31	23	74.19	32	23½	73.43	
Mexican	1834	M.	26	21	80.76	26	21½	82.69	Youth.
Mahlemut Eskimo	756	....	30	20	66.66	29	20½	70.68	
North American Indians.									
Alaskan	814	M.	38½	20½	53.24	.....	.....	.....	
Do	938	M.	34	23	67.64	34½	22½	65.21	Hunchback.
Apache	1473	F.	27	18½	68.51	29½	20	67.79	
Do	1530	M.	34	24	70.58	33	23½	71.21	
Arapaho	6489	M.	37	23	62.16	38½	24½	63.63	
Bannock	2123	M.	33½	28½	85.07	32	30	93.75	
Cheyenne	2036	M.	32½	23	70.76	33½	23	68.65	
Do	2066	M.	41	25	60.97	42	23	54.76	
Do	6866	M.	34½	24½	71.01	35	25	71.42	Adolescent.
Chippewa	1257	M.	33	22	66.66	34	23½	69.11	
Choctaw	622	F.	32½	20	61.53	31½	21½	68.25	
Do	623	F.	31½	21	66.66	30½	21	68.85	
Comanche	1000	F.	28	19	67.85	28	19	67.85	
Dakota	945	M.	40½	27½	67.90	40½	27	66.66	
Modoc	6287	M.	35	23½	67.14	35	22	62.85	
Mound-builders:									
From Dakota	165	M.	38½	24½	63.63	38½	25	64.93	
Do	1121	M.	42	24½	58.33	39½	24½	62.02	
From Mississippi	400	M.	42½	30	70.58	39½	27½	69.62	
Do	644	M.	33	24	72.72	34½	25	72.46	
Navajo	788	M.	33	21½	65.15	34	22½	66.17	Right foramen abnormal; measurement taken on a level to correspond with foramen of opposite side.
Pawnee	778	F.	32	23½	73.43	31½	21½	68.25	
Pah Ute	963	M.	35	21	60.00	35	22	62.85	
Do	964	F.	30	17	56.66	28½	17	59.64	
Sioux	13	M.	36½	24½	67.12	36	25	69.44	
Do	2046	F.	31	24	77.41	31	24	77.41	
Do	2071	M.	34	23½	69.11	35	23½	67.14	
Do	2072	F.	32½	24	73.84	33½	23½	70.14	
Brule Sioux	1892	M.	34	25	73.52	34½	23	66.66	Left foramen abnormal; measurement taken on a level to correspond with foramen of opposite side.
Do	1895	M.	34½	24	69.56	36	23	63.88	
Do	1896	M.	36½	27	73.97	37	29	78.37	
Do	1897	F.	35	27	77.14	36	25½	70.83	
Do	1898	M.	41	28	68.29	41	27	65.85	
Do	1899	M.	43	23	53.48	42½	23½	55.29	
Do	1900	F.	30	21½	71.66	30	21	70.00	36 years of age.
Do	1901	M.	38½	24½	63.63	38	24	63.15	
Do	2061	M.	35½	24½	69.01	35	23½	67.14	Adolescent.
Do	2062	M.	41½	26½	63.85	38½	27½	71.42	
Do	2063	M.	28	21½	76.78	29½	22	74.57	Adolescent.
Do	2068	F.	36	24½	68.05	33	24½	74.24	
Ogalalla Sioux	1851	F.	27	21	77.77	27	20½	75.92	About 15 years of age.
Do	1853	F.	31	20½	66.12	32	20½	64.06	Left foramen abnormal; measurement taken on a level to correspond with foramen on opposite side.
Do	1854	F.	32½	21½	66.15	33	21½	65.15	
Do	1856	M.	26½	19½	73.58	27	20	74.0	About 16 years of age.
Sisseton Sioux	1790	F.	30	22	73.33	30½	22	72.13	
Do	1791	F.	40	23	57.50	38½	24	62.33	
Do	1792	M.	38	28½	75.00	38½	29	75.32	
Yankton Sioux	936	M.	39½	25	63.29	36	25	69.44	

TABLE LXXVII—Continued.

Races.	No. of specimen.	Sex.	Right tibia.			Left tibia.			Remarks.
			Antero posterior dimension.	Lateral dimension.	Index.	Antero posterior dimension.	Lateral dimension.	Index.	
South American Indians.									
Patagonian	239	F.	36	17	58.66	31	17	54.83	
Peruvian	1595	F.	28½	18½	64.91	28	19	67.85	
Asiatics.									
Chinese	956	M.	37	29	78.37	36	27	75.00	*

TABLE LXXVIII.—*Synopsis of average indices of section of the femur and of section of the tibia in certain numbers of skeletons in the Army Medical Museum.*

Races.	Average indices of section of femur.				Average indices of section of tibia.			
	No. of femurs.	Right side.	No. of femurs.	Left side.	No. of tibiæ.	Right side.	No. of tibiæ.	Left side.
Sioux Indians	24	112.48	24	110.33	24	69.54	24	69.33
Other North American Indians	23	113.60	21	111.89	23	66.44	22	67.54
Negroes	6	120.53	6	119.10	6	75.00	6	75.00

The **following** formulæ are found used in the various articles on craniology in the *Zeitschrift für Ethnologie*, from 1879 to 1889, inclusive, to reckon various facial indices. A few articles concerning very small numbers of skulls are omitted. The page given is that on which the article begins. The articles sometimes are made up by two or three men, but Virchow generally writes the craniometrical part.

In these articles (rejecting two articles where, if the formulæ indicated are correct, gross arithmetical errors have been made; also two where the formulæ have been reversed—a clerical error perhaps—and the translation of the article in Italian by Raf. Zampa, where the terminology is a little suspicious) we have the various formulæ occurring as follows:

TABLE LXXIX.

Year.	Page.	Title.	Author.	Formula.
1879	118	Livländische Schädel	Virchow	Gesichtshöhe × 100. Gesichtsbreite (b) Bizygom and Obergesichtshöhe × 100. Gesichtsbreite (b) Bizygom.
1879	136	Ueber Schädel von Ophrynium	Virchow	Gesichtshöhe × 100. Gesichtsbreite, Sut. zyg. max and Obergesichtshöhe × 100. Gesichtsbreite, Sut. zyg. max
1879	422	Vier Schädel von Cagrasay (Philippinen)	Virchow	Mittelgesichtshöhe × 100. Gesichtsbreite, Malar and Mittelgesichtshöhe × 100. Gesichtsbreite, jugal.
1880	53	Höhlenschädel aus dem oberen Weichselgebiet	Virchow	Gesichtshöhe (b Alveolarrand) × 100. Malarbreite (Bizygom).
1880	121	Schädel von Tehn und Westafrikanern	Virchow	Breite des Gesichts × 100. Höhe des Gesichts.

TABLE LXXIX—Continued.

Year.	Page.	Title.	Author.	Formula.
1881	226	Schädel von Madisonville, Ohio, und von Casabamba, Süd-Colombien.	Virchow	$\dfrac{\text{Gesichtshöhe (a)} \times 100}{\text{Jochbreite}}$
1881	357	Das Gräberfeld von Slaboszewo bei Mogilno	Virchow	$\dfrac{\text{Gesichtshöhe A} \times 100}{\text{Gesichtsbreite A (jugal)}}$ and $\dfrac{\text{Gesichtshöhe B} \times 100}{\text{Gesichtsbreite A (jugal)}}$
1882	76	Alfuren-Schädel von Ceram und anderen Molucken	Virchow	$\dfrac{\text{Gesichtshöhe (A)} \times 100}{\text{Jugalbreite}}$ and $\dfrac{\text{Mittelgesichtshöhe (B)} \times 100}{\text{Jugalbreite}}$
1882	298	Die Weddas auf Ceylon	Virchow	$\dfrac{\text{Mittelgesichtshöhe} \times 100}{\text{Jugalbreite}}$
1883	306	Die Rasse von La Tène	Virchow	$\dfrac{\text{Gesichtshöhe A} \times 100}{\text{Gesichtsbreite A}}$
1883	331	Eine Fibula aus der Tschetschna und zwei Schädel von Koban.	Virchow	$\dfrac{\text{Gesichtshöhe B} \times 100}{\text{Gesichtsbreite b, malar.}}$ (N. B.—But both indices are wrong if this is the correct formula.)
1883	390	Schädel der Igorroten	Virchow	$\dfrac{\text{Gesichtshöhe A} \times 100}{\text{Gesichtsbreite a, jugal,}}$ and $\dfrac{\text{Gesichtshöhe B} \times 100}{\text{Gesichtsbreite b, malar.}}$
1884	181	Hohes Alter der Menschenrassen	Kollmann	$\dfrac{\text{Gesichtshöhe} \times 100}{\text{Jochbreite}}$ and $\dfrac{\text{Obergesichtshöhe} \times 100}{\text{Jochbreite.}}$
1884	47	Burgwall bei Ketzin	Virchow	$\dfrac{\text{Gesichtshöhe A} \times 100}{\text{Gesichtsbreite A.}}$
1884	115	Das neolithische Gräberfeld von Tangermünde	Virchow	$\dfrac{\text{a, Gesichtshöhe A} \times 100}{\text{Gesichtsbreite, a, jugal}}$ and b, doubtful.
1884	168	Die Rasse von La Tène	Virchow	$\dfrac{\text{Gesichtshöhe A} \times 100}{\text{Gesichtsbreite a, jugal.}}$
1884	308	Schädel mit zwei Schläfenringen aus Nakel	Virchow	$\dfrac{\text{Gesichtshöhe A} \times 100}{\text{Gesichtsbreite A.}}$
1884	355	Gräberfelder und Urnenfaule bei Tangermünde	Virchow	$\dfrac{\text{Gesichtshöhe} \times 100}{\text{Gesichtsbreite a, jugal.}}$
1884	390	Anthropologische Excursion nach Feldberg	Virchow	$\dfrac{\text{Gesichtshöhe B} \times 100}{\text{Gesichtsbreite B, malar.}}$
1885	45	Die Bewohner von Süd-Mindanao und der Insel Samal.	Schadenberg.	$\dfrac{\text{Gesichtshöhe} \times 100}{\text{Gesichtsbreite (Sut. zyg. max.}}$ and $\dfrac{\text{Obergesichtshöhe} \times 100}{\text{Gesichtsbreite (Sut. zyg. max.}}$ and $\dfrac{\text{Gesichtshöhe} \times 100}{\text{Jochbreite}}$ and $\dfrac{\text{Obergesichtshöhe} \times 100}{\text{Jochbreite.}}$
1885	248	Schädel und Skelette von Botocuden am Rio Doce.	Virchow	$\dfrac{\text{Gesichtshöhe A} \times 100}{\text{Gesichtsbreite a, jugal.}}$

TABLE LXXIX—Continued.

Year.	Page.	Title.	Author.	Formula.
1885	283	Pfahlbauschädel des Museums in Bern	Virchow	Gesichtshöhe B × 100 / Gesichtsbreite b, malar.
1885	314	Reise nach Angra Pequena und Damaraland	Virchow	Gesichtsbreite B malar × 100 / Gesichtshöhe A. (N. B.—Formula reversed.)
1885	497	Wedda-Schädel	Virchow	Gesichtshöhe B × 100 / Gesichtsbreite b, malar.
1886	205	Vergleichende anthropologische Ethnographie von Apulien.	Zampa	Gesichtslänge × 100 / Gesichtsbreite and Ganze Gesichtslänge × 100 / Breite der Wangenpunkte. (N. B.—Gesichtslänge apparently means both upper and entire facial height.)
1886	429	Ausgrabungen an der Insel im See von Jankowo.	Virchow	Gesichtshöhe × 100 / Gesichtsbreite (jugal).
1886	652	Ein Skelet und Schädel von Guajiros	Virchow	Gesichtshöhe A × 100 / Gesichtsbreite A.
1886	732	Schädel von Baluba und Congonegern	Virchow	Gesichtshöhe A × 100 / Gesichtsbreite A and Gesichtshöhe B × 100 / Gesichtsbreite B.
1887	49	Ueber die Botocudos	Ehrenreich	Obere Gesichtshöhe × 100 / Jochbreite and Obere Gesichtshöhe × 100 / Gesichtsbreite.
1887	296	Motilonen Schädel aus Venezuela	Ernst	Gesichtshöhe A × 100 / Gesichtsbreite a, jugal and Gesichtshöhe B × 100 / Gesichtsbreite b, malar. (N. B.—Indices wrongly figured.)
1887	321	Gräberfunde von den Key-Inseln, Molukken	Virchow	Gesichtshöhe A × 100 / Gesichtsbreite A.
1887	354	Gräberfund von Kawenczyn, Posen	Virchow	Gesichtshöhe A × 100 / Gesichtsbreite A and Gesichtshöhe b × 100 / Gesichtsbreite b.
1887	451	Schädel von Merida, Yucatan	Virchow	Gesichtshöhe B × 100 / Gesichtsbreite b (malar).
1887	624	Anthropologie der Völker vom mittleren Congo	Mense	Gesichtshöhe a × 100 / Gesichtsbreite a and Gesichtshöhe b × 100 / Gesichtsbreite b.
1888	578	Siamesen-Schädel	Virchow	Gesichtshöhe A × 100 / Gesichtsbreite a.
1889	170	Schädel von Tenimber und Letti (Timor), Sunda-Inseln.	Virchow	Gesichtshöhe A × 100 / Gesichtsbreite a and Gesichtshöhe B × 100 / Gesichtsbreite a.

TABLE LXXIXa.—*Summary showing frequency of use of the various German indices in table LXXIX.*

Name of index from "Verständigung."	Formulae.	No. of crossing.
Facial index of Kollmann	Mento-nasal height × 100 / Bizygomatic width	22
Upper facial index of Kollmann	Alveolo-nasal height × 100 / Bizygomatic width	8
Facial index of Virchow	Mento-nasal height × 100 / Biumaxillary width	4
Upper facial index of Virchow	Alveolo-nasal height × 100 / Biumaxillary width	14

TABLE LXXIXb.—*Facial indices of Virchow from Europeans.*

Facial	119.1	Upper facial	73.1
Average of	11.0	Average of	27.0

TABLE LXXX.—*Craniometrical data according to Frankfurt agreement, computed from data quoted in Table LXXIX.*

Races.	Length-breadth index. Aver. age.	No. of skulls.	Length-height index. Aver. age.	No. of skulls.	Facial index of Kollmann. Aver. age.	No. of skulls.	Upper facial index of Kollmann. Aver. age.	No. of skulls.	Facial index of Virchow. Aver. age.	No. of skulls.	Upper facial index of Virchow. Aver. age.	No. of skulls.	Nasal index. Aver. age.	No. of skulls.	Palatine index. Aver. age.	No. of skulls.	Facial angle. Aver. age.	No. of skulls.
Malays	77.2	28	77.0	25	84.3	8	48.6	16	111.5	3	67.9	24	53.2	28	72.1	22		
Veddahs	76.8	5	78.2	5			53.8	3										
Negroes	75.9	21	78.1	12	95.4	6					67.8	15	54.9	19	69.2	17	69·12	10
Botocudos	74.5	17	76.2	17	89.1	3	52.6	11			70.6	11	47.9	18	71.3	3	80·750	10
Guajiros skulls	81.2	8	73.8	8	83.3	8							47.2	8	72.5	7		
Motilo skull	79.0		75.6		80.0						66.0		48.8		78.0			
Yucatan skull	90.2		75.7								65.7		49.0		80.3			
Calaveras skull					76.3		42.6						58.6		100.0			
Rock Bluff skull	74.0		71.5		86.5		47.0						50.0		84.8			
Lagoa Santa skulls	72.2	5	80.2	5	84.2	4	47.0	4					50.2	5	98.3			

TABLE LXXXI.—*Special series of 101 skulls in the general collection of the Army Medical Museum.*

NOTE.—Frequent reference is made in this work to our series of 101. This is a collection of 101 adult skulls, representing 27 different tribes and races, which we measured exactly on the same system that the Hemenway collection was measured, just previous to commencing the study of the latter. Although it is a small series, we have found it useful for making comparisons in preparing this essay.

The composition of the series of 101 is as follows:

Race or tribe.	Skulls of females.	Skulls of males.	Skulls of unknown sex.	Total.	Race or tribe.	Skulls of females.	Skulls of males.	Skulls of unknown sex.	Total.
Pah Utes	3	4		7	Chuckchees		2		2
Apaches	3	3		6	Japanese	1	1		2
Ancient Californians	5	5		10	Chinese	1	1		2
Cheyennes	1	1		2	Australians	1	1	1	3
Chippewas	1	1		2	Fiji Islanders	1	1		2
Minnetarees	1	1		2	Sandwich Islanders	3	3		6
Navajos	2	2		4	New Zealanders	2	2		4
Pawnees		2		2	Chatham Islanders	1	1		2
Poncas	2	2		4	American Negroes	3	3		6
Seminoles		2		2	American Whites	3	3		6
Sioux	2	2		4	Bavarians	1	1		2
Yucatecs	1	1		2	Austrians	3	3		6
Eskimos, Alaskan	1	3		4					
Eskimos, Greenlandic	3	3		6	Total	46	54	1	101
Eskimos, Asiatic		1		1					

TABLE LXXXII.—*General measurements.—Cibola.*

Number of skull	Greatest length	Greatest width	Basibregmatic height	Basialveolar diam.	Basi-nasal radius.	Mesio-nasal height	Interorbitary width	Nasal height	Nasal width	Orbital width	Orbital height	Prostic angle	Angle of Prohibition	Occipital angle	Basilar angle
H. 201	174	132	133	92	97	......	......	44	25	38	31	89	.	.	.
H. 202	165	152	151	95	102	121	108	49	24	43	34	88½	4½	14	17
H. 203	164	155	146	102	101	123	107	51	26	41	34	85	9	20½	26
H. 204	185	140	128	......	103	......	......	......	......	40	33	......	9	17	22
H. 205	145	140	133	92	92	109	103	49	25	38	34	84	6	18½	24
H. 206	146	144	138	93	94	114	96	50	25	39	36	83	5	15	20
H. 207	165	131	......	......	......	98	95	45	26	39	35	81	......	......	......
H. 208	160	130	......	......	......	104	88	51	23	37	35	87	......	......	......
H. 209	165	123	133	96	97	106	93	47	25	37	34	82	6½	18	24
H. 210	146	142	133	90	90	107	91	44	23	36	34	86½	4	16	21½
H. 211	167	145	147	108	105	124	113	52	28	41	35	85	......	......	......
H. 212	155	146	136	95	95	......	88	47	26	38	34	80	2	13	18
M. 213	170	142	141	93	101	118	106	48	27	39	36	88	10½	22½	30
H. 214	163	145	......	......	......	109	101	49	26	38	33	88½	......	......	......
H. 215	157	135	135	98	99	112	101	51	25	41	35	87	......	......	......
H. 216	143	141	145	91	95	......	95	48	24	38	32	88	6	17	23
H. 217	152	146	147	......	99	......	100	45	27	40	35	......	8½	18½	25
H. 218	174	151	147	97	103	......	107	54	28	40	35	......	7	18½	24½
H. 219	163	141	144	97	101	......	88	46	23	36	33	85	......	......	......
H. 220	151	142	138	99	96	118	95	49	23	38	35	85	8½	20½	27½
H. 221	163	133	131	98	96	112	99	50	25	36	34	83	1	10½	14
H. 222	151	134	138	89	97	106	99	47	26	39	36	89	11	22	30
H. 223	166	145	138	97	98	109	94	48	25	37	33	86	1½	11	15
H. 224	148	141	145	100	100	119	101	50	26	40	35	81	13	24½	32
H. 225	152	139	137	......	......	112	91	47	24	40	34	83	10½	22	29
H. 226	167	144	145	96	102	120	101	51	24	36	33	91	10	20½	26½
H. 227	147	139	136	88	95	102	100	48	24	37	34	89	4	14½	19½
H. 228	157	146	146	95	97	119	106	50	28	40	36	84	2½	13	17
H. 229	166	129	132	90	96	118	95	51	23	38	35	91½	2	13½	18
H. 230	157	144	134	97	98	104	104	51	26	38	34	85	10½	20½	27
H. 231	162	159	......	......	......	104	92	42	23	37	34	81	......	......	......
H. 232	159	140	137	95	97	118	103	50	28	39	33	84	11	21½	29
H. 233	163	150	144	99	100	122	101	54	25	38	30	88½	5½	16	22
H. 234	159	152	146	95	95	111	97	48	25	35	32	88	7	17	22
H. 235	147	140	138	89	96	109	94	46	24	39	34	86	10	23	31

TABLE LXXXIII.—*Indices of 35 skulls.—Cibola.*

No. of skull.	Cephalic.	Vertico-longitudinal.	Gnathic.	Facial of Virchow.	Nasal.	Orbital.
H. 201	75. 86	76. 43	94. 84	....	56. 81	81. 57
H. 202	92. 12	91. 31	93. 13	112. 03	48. 16	79. 06
H. 203	94. 51	89. 62	100. 99	111. 95	50. 38	82. 92
H. 204	75. 67	74. 05	....	....	.....	82. 50
H. 205	96. 55	91. 72	100. 00	105. 82	51. 02	89. 47
H. 206	98. 63	94. 92	98. 93	108. 75	50. 00	92. 30
H. 207	80. 36	....	....	103. 15	60. 46	89. 74
H. 208	76. 92	....	....	123. 89	45. 09	94. 59
H. 209	74. 51	89. 66	98. 96	113. 97	52. 19	94. 89
H. 210	97. 26	91. 09	100. 00	117. 58	52. 27	94. 44
H. 211	86. 82	88. 62	102. 85	109. 73	53. 84	85. 36
H. 212	94. 19	87. 74	100. 00	....	55. 31	85. 47
H. 213	88. 62	82. 94	92. 07	111. 32	56. 25	92. 30
H. 214	88. 95	....	....	107. 92	53. 06	85. 84
H. 215	85. 98	85. 98	98. 98	110. 89	49. 01	85. 36
H. 216	100. 69	101. 39	95. 78	....	50. 00	84. 21
H. 217	96. 05	95. 71	....	....	60. 00	87. 50
H. 218	80. 78	81. 48	94. 17	....	51. 85	87. 50
H. 219	86. 50	88. 51	96. 63	....	50. 00	91. 66
H. 220	94. 02	91. 39	103. 12	124. 21	46. 93	92. 10
H. 221	81. 39	80. 35	102. 08	115. 13	50. 00	94. 44
H. 222	88. 74	91. 39	91. 75	107. 07	55. 31	92. 30
H. 223	87. 31	83. 13	98. 97	115. 95	52. 90	89. 18
H. 224	95. 27	97. 97	100. 00	117. 82	52. 00	87. 50
H. 225	91. 44	90. 13	....	123. 07	51. 06	85. 00
H. 226	86. 22	86. 82	94. 11	118. 81	47. 05	91. 66
H. 227	94. 55	92. 51	92. 63	102. 00	50. 00	91. 89
H. 228	92. 90	92. 99	97. 93	112. 26	56. 60	90. 00
H. 229	77. 71	79. 54	93. 75	124. 21	45. 09	92. 10
H. 230	91. 71	85. 35	98. 97	109. 61	50. 98	89. 47
H. 231	85. 80	....	....	109. 78	51. 75	91. 89
H. 232	88. 05	90. 16	97. 93	114. 56	56. 60	81. 61
H. 233	92. 02	88. 34	99. 00	120. 79	46. 29	78. 94
H. 234	95. 59	91. 82	100. 00	114. 43	52. 08	91. 42
H. 235	95. 23	93. 88	92. 70	112. 76	52. 17	87. 17

TABLE LXXXIV.—*Summary of indices of skulls.—Cibola.*

Indices.	Number of skulls.	Maximum.	Minimum.	Average.
Cephalic	35	100. 69	74. 51	88. 86
Vertico-longitudinal	31	101. 39	74. 05	88. 28
Gnathic	28	103. 12	91. 75	97. 48
Facial of Virchow	28	124. 21	102. 00	113. 91
Nasal	34	60. 46	45. 09	51. 88
Orbital	35	94. 59	78. 94	88. 52

H. 220 and H. 229 both show maximum *facial* indices.
H. 208 and H. 229 both show minimum *nasal* indices.

TABLE LXXXV.—*Summary of angles of skulls.—Cibola.*

Angles.	Number of skulls.	Maximum.	Minimum.	Average.
		○	○	○
Profile angle	32	91½	80	85½+
Angle of Daubenton	27	13	1	6½+
Occipital angle	27	24½	10½	17½+
Basilar angle	27	32	14	23½

TABLE LXXXVI.—*Classification of the inion.—Cibola.*

Serial number.	No. of skull.	Class.	Serial number.	No. of skull.	Class.	Serial number.	No. of skull.	
1	H. 201	1	12	H. 213	1	23	H. 225	1
2	H. 202	3	13	H. 215	2	24	H. 226	3
3	H. 203	1	14	H. 216	1	25	H. 227	0-1
4	H. 204	2	15	H. 217	1	26	H. 228	1
5	H. 205	0-1	16	H. 218	4 (?)	27	H. 229	1-2
6	H. 206	0	17	H. 219	2-3	28	H. 230	1
7	H. 208	0	18	H. 220	0-1	29	H. 232	1
8	H. 209	0	19	H. 221	2-3	30	H. 233	1-2
9	H. 210	0	20	H. 222	2	31	H. 234	0
10	H. 211	3-4	21	H. 223	3	32	H. 235	0-1
11	H. 212	1	22	H. 224	1			

In Nos. H. 205, H. 208, H. 209, **H.** 212, H. 215, H. 220, H. 225, and **H. 226** the inion is less prominent than adjoining points on the superior curved line.

TABLE LXXXVII.—*Seriation of the inion in 32 skulls.—Cibola.*

0	5	3		3
0-1	4	3-4		1
1	11	4		1
1-2	2	5		0
2	3			
2-3	2	Total		32

TABLE LXXXVIII.—*Pteria of skulls.—Cibola.*

No. of skull.	Length of right pterion.	Length of left pterion.	No. of skull.	Length of right pterion.	Length of left pterion.
H. 202	14	12	H. 216		7
H. 203	19	21	H. 221	15	12
H. 205	15	17	H. 222		19
H. 206	10	7	H. 223	15	11
H. 207	18		H. 226	10	9
H. 208	15	9	H. 227	13	14
H. 209	15	13	H. 228	17	17
H. 210	17		H. 229	17	16
H. 211	17	19	H. 230	9	13
H. 212	12	11	H. 231	8	11
H. 213	17	16	H. 234	18	14

Average of right pterion = 14.60^{mm}; average of left pterion = 13.55^{mm}. Mean average = 14.07^{mm}. H. 203 has epipteric bone on right side. H. 230 has epipteric bones right and left.

TABLE LXXXIX.—*Showing the character of the Échancrure, or lower nasal border, in 34 skulls.—Cibola.*

Class (Broca).	No. of skulls.	Percentage.	Designation of the skulls.
A	4	11.76	H. 201, H. 218, H. 220, H. 229.
A¹	9	26.47	H. 202, H. 206, H. 209, H. 212, H. 214, H. 215, H. 222, H. 231, H. 234.
A + A¹	13	38.23	
B	12	35.27	H. 205, H. 207, H. 208, H. 211, H. 213, H. 216, H. 221, H. 225, H. 226, H. 227, H. 230, H. 235.
C	6	17.64	H. 203, H. 210, H. 217, H. 223, H. 224, H. 228.
D	3	8.82	H. 219, H. 232, H. 233.
E	0	0.00	

TABLE XC.—*Ordination of the angles of torsion of the humeri arranged according to the left humeri.—Cibola.*

Designation	Left humeri		Right humeri		Designation	Left humeri		Right humeri	
	o	'	o	'		o	'	o	'
1 H. 233	143	00	145	00	13 H. 221	160	00	148	00
2 H. 209	149	00	144	00	14 H. 224	160	00	149	00
3 H. 214	150	30	....		15 H. 216	161	00	158	30
4 H. 207	151	00	136	30	16 H. 227	161	00	157	00
5 H. 213	152	30	137	00	17 H. 226	163	00	146	30
6 H. 217	153	00	....		18 H. 212	165	00	162	30
7 H. 215	154	00	137	00	19 H. 211	170	00	151	00
8 H. 206	154	30	150	00	20 H. 218	170	00	148	30
9 H. 234	155	00	152	00	21 H. 228	170	30	....	
10 H. 220	157	00	146	00	22 H. 229	171	00	154	00
11 H. 204	157	30	153	00	23 H. 222	178	00	165	00
12 H. 205	158	30	151	00					

Average angle of left humerus = 159° 20'.

TABLE XCI.—*Ordination of the angles of torsion of the humeri arranged according to the right humeri.—Cibola.*

Designation	Right humerus		Left humerus		Designation	Right humerus		Left humerus	
	o	'	o	'		o	'	o	'
1 H. 207	136	30	151	00	14 H. 206	150	00	154	30
2 H. 213	137	00	152	30	15 H. 205	151	00	158	30
3 H. 215	137	00	154	00	16 H. 211	151	00	170	00
4 H. 203	143	00	....		17 H. 234	152	00	155	00
5 H. 209	144	00	149	00	18 H. 204	153	00	157	30
6 H. 233	145	00	143	00	19 H. 202	154	00	....	
7 H. 223	145	30	....		20 H. 229	154	00	171	00
8 H. 232	145	30	....		21 H. 227	157	00	161	00
9 H. 220	146	00	157	00	22 H. 216	158	30	161	00
10 H. 226	146	30	163	00	23 H. 208	162	30	....	
11 H. 221	148	00	160	00	24 H. 212	162	30	165	00
12 H. 218	148	30	170	00	25 H. 222	165	00	178	00
13 H. 224	149	00	160	00					

Average angle of right humerus = 149° 40'.

APPENDIX A.

Those which we use, only, are here given.

Essential measurements.

1. Greatest antero-posterior length: From the glabella to the maximum occipital point.
2. Greatest transverse width: Upon the parietal or squamous portion of temporal, no matter where the maximum may fall.
3. Basilo-bregmatic diameter: From the basion to bregma.
4. Smallest frontal width: Shortest distance between the temporal ridges of the frontal bone.
5. Horizontal circumference: Horizontal circumference of the cranium directly above the superciliary ridge and across the most prominent point of the occiput.
7. Naso-basilar line: Nasion to basion.
8. Maximum bizygomatic width: Greatest distance between the zygomatic arches.
9. Biorbital width: Maximum external biorbital or bimalar width from external extremity of small fronto-malar suture to same point opposite.
11. Maximum bimaxillary width: Maximum distance between the inferior extremity of the maxillo-malar suture to the corresponding opposite point.
12. Bigonial width: From the external portion of one angle of the jaw to another.
17. Nasal height or naso-spinal height: From the nasion to the middle of the upper border of the lower nasal spine or lower border of nasal aperture.
18. Maximum width of nasal aperture.
19. Width of orbit: From the dacryon to the opposite external margin following the direction of the grand axis.
20. Height of orbit: Perpendicular to the preceding, beginning at middle of inferior border.
22. Occipito-alveolar length: From the maximum occipital point to the alveolar point.
23. Occipito-spinal length: From the maximum occipital point to the inferior border of the nasal aperture.
24. Capacity of the cranium: Broca's method.

Complimentary measurements.

A. Antero-posterior metopic length: From the metopion to the maximum point of the occiput.
B. Biasteric or maximum occipital width.
C. Bijugular or inferior occipital width.
E. Bitemporal width: From one subtemporal point to another.
F. Vertical circumference or supra-auricular curve: Between the two supra-auricular points, passing upon the bregma.
G. Anterior and posterior parts of horizontal circumference separated by the supra-auricular curve.
H. Interorbital width: Distance from one dacryon to the other.
I. Alveolar external maximum width: Taken at the level of the molar region.

---

[*] TOPINARD: Éléments d'Anthropologie Générale, Paris, 1885, pp. 919, et seq.

J. Alveolar external posterior width: Taken at the junction of the exterior arch and the *pan coupé* which is behind the wisdom tooth beneath the articulation of the pterygoid apophysis.

K. Anterior palatine width: Taken between canine and second incisor.

M. Posterior palatine width.

P. Palatine depth.

Q. Height and width of the posterior branch of mandible: Height from angle to upper edge of condyle; width at right angle with the above.

R. External bicondylar width: Taken between the outer edges of the condyles of mandible.

S. Basilo-mental radius.

U. Superior alveolar radius.

V. Nasal radius.

W. Intersuperciliary radius.

X. Metopic radius.

Y. Obelic radius.

Z. Inial radius.

d. Anterior projection of the cranium or pre-basilar projection.

e. Posterior projection of the cranium or post basilar projection.

f. Superior facial projection or projection of the ophryon.

     The above three (d, e, and f) should be taken with regard to alveolo-condylean plane.

APPENDIX B.

CRANIOMETRICAL DIRECTIONS OF THE FRANKFURT AGREEMENT,* AUGUST, 1882.

Those which we use are here given in full, with the numbers, reference letters, etc., which they bear in the original work.

Horizontal plane of cranium indicated by two lines connecting the lowest points of the borders of the orbits with the upper margins of the *meati auditorii* at points perpendicularly above their centers.

1. Horizontal length from the central point between the superciliary ridges to the most prominent part of the occiput on a level with the horizontal plane measured with calipers. The *horizontal length* (L, Figs. 1 and 2) is measured parallel to the horizontal plane, and the taking of this measurement shall be by means of the sliding calipers or Spengel's craniometer. Why this is necessary is significantly shown in Fig. 2. If one takes this measurement with the ordinary calipers, especially on a very long skull with strongly projecting occiput, the result is too small if the measurement is not continued to the tangent, which, rising vertically from the horizontal plane, touches the farthest point of the occiput. This can be done only

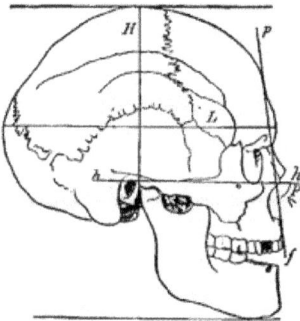 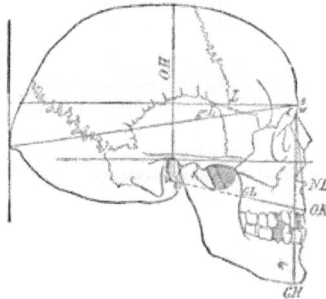

FIG. 1.   FIG. 2.

with one of the above-mentioned instruments. Indeed, even in their use experience is necessary, and repeated control-experiments. In skulls with full round occiputs the taking of this measurement has no difficulties. As Fig. 1 shows, the most projecting point lies in the same height as the anterior end of L. Respecting this latter point on the glabella (marked S in Fig. 2) a mistake is impossible. Always put the measuring instrument in the median line, therefore, between the superciliary ridges whenever they are separated. As to the *greatest length* (gr. L, Fig. 2), it is apparent, upon a comparison of Figs. 1 and 2, that a difference between this and the horizontal length can only occur in skulls with very prominent occiputs. In the full, rounded occiput of Fig. 1 both lengths are identical. The sliding calipers and ordinary calipers, accurately applied, give then the same result. In the extreme case taken in Fig. 2 the difference, with a length of 206ᵐᵐ for the brain capsule, amounts to 5ᵐᵐ. Also, the skull length measured from the line of the frontal protuberance, the *intertuberal length*, coincides in its results very nearly with the greatest length and horizontal length, especially in brachicephalic skulls with well-rounded foreheads.

---

* Archiv für Anthropologie, Bd. xv, 1884, pp. 1–8.

2. Greatest length (longitudinal diameter) from the center point between the superciliary ridges to the most prominent part of the occiput (without regard to horizontal plane); calipers.

3. Intertubal length from the central point between the frontal eminences and the most prominent part of the occiput (without regard to the horizontal plane); measured with calipers.

4. Greatest width, B. B., Fig. 3, perpendicular to sagittal plane, measured with calipers (not over the mastoid processes or at the posterior temporal ridge); the points measured must be on the same horizontal plane.

5. Smallest frontal breadth, S. S., Fig. 4; shortest distance between the temporal ridges of the frontal bone.

6. Height, called entire height after Virchow, II, Fig. 1: From the center of the anterior border of the foramen magnum to the parietal curve, perpendicular to horizontal plane. The difference between the height of the posterior border of the foramen magnum and the anterior should be indicated from which the height according to Baer-Ecker is ascertained. (Measured with calipers.)

7. Auxiliary height: As in crania, in which the bones of the face are missing, the horizontal plane can not be accurately indicated, the following shall be measured as the auxiliary height: From the center of the anterior border of the foramen magnum to junction of coronal and sagittal sutures; this always nearly corresponds with the height as in 6.

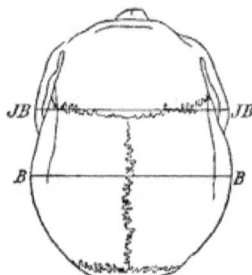
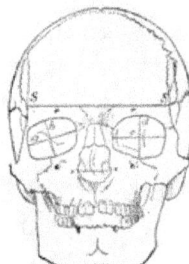

Fig. 3.   Fig. 4.

8. Auricular height: From the upper margin of the meatus auditorius to a point of the vertex perpendicularly above the meatus, perpendicular to the horizontal plane.

9. Auxiliary auricular height: From the same starting point to the highest point of the parietal curve, about 2 or 3 centimeters behind the coronal suture.

10. Length of cranial basis: From the center of the anterior border of the foramen magnum to the middle of the naso-frontal suture. (Measured with calipers.)

12 and 13. Greatest length and breadth of foramen magnum to be measured in the sagittal plane and perpendicular thereto.

13a. Breadth of cranial basis: Distance between the ends of the mastoid processes.

14. Horizontal circumference of cranium: Directly above the superciliary ridge and over the most prominent part of the occiput. Steel-tape.

15. Sagittal circumference of cranium: From the naso-frontal suture to the posterior margin of the foramen magnum along the sagittal suture. Steel-tape.

16. Vertical circumference from one upper margin of the meatus auditorius to the other, perpendicular to horizontal plane (about 2 or 3 centimeters behind coronal suture). Steel-tape.

LINEAL MEASURES OF FACE.

17. Facial width after Virchow: Distance between the maxillo-malar sutures; should be measured from the lower anterior corner of one malar bone to the other.

18. Zygomatic width: Greatest distance between the zygomatic arches.

18a. Interorbital width: Shortest distance between the inner borders of the orbits.

19. Facial height from the center of the fronto-nasal suture to the center of the lower border of the inferior maxilla. (G. H.—W, Fig. 2.)

20. Upper (or middle) facial height: From the middle of the naso-frontal suture to the middle of the alveolar edge of the superior maxilla, between the middle incisors. (O. K.—W, Fig. 2.)

21. Nasal height (W.—N. H., Fig. 2): From the middle of the naso-frontal suture to the middle of the upper border of the lower nasal spine.

22. Greatest breadth of nasal cavity (wherever it is found, see Fig. 4) to be measured horizontally.

23. Greatest breadth of orbit (a Fig. 4): From middle of median border to lateral border of orbit.

25. Greatest height of orbit (Fig. 4, b): Perpendicular to greatest breadth.

27. Length of palate bone: From the extreme point of the posterior nasal spine to the inner lamella of the alveolar border between the middle incisors.

28. Median width of palate: Between the inner alveolar walls of the second molars.

29. Width of posterior end of palate: On both posterior ends of palate, between the inner alveolar walls.

30. Length of profile of face) Kollmann's (G. L., Fig. 2): From the most prominent part of the middle of the external alveolar border of the upper maxilla to the anterior margin of the foramen magnum (in the median plane).

31. Profile angle (P <, Fig. 1) is the angle formed by profile line Pf with the horizontal.

MEASUREMENT OF CAPACITY OF CRANIUM.

32. The capacity of the cranium is measured with shot (in fragile crania with millet). The manner of measuring to be agreed upon hereafter.

CRANIAL INDICES.

I. Length-width index $\frac{100 \text{ width}}{\text{length}}$.

Dolichocephalic	to 75.0
Mesocephalic	75.1—79.9
Brachycephalic	80.0—85.0
Hyperbrachycephalic	from 85.1 and over.

II. Length-height index $\frac{100 \text{ height}}{\text{length}}$.

Chamæcephalic (flat crania)	to 70.0
Orthocephalic	70.1—75.0
Hypsicephalic (high crania)	75.1 and over.

III. Profile angle.

The inclination of the profile line to the horizontal is divided in the following three grades:

1. Prognathic	to 82.0
2. Mesognathic or orthognathic	83.0—90.0
3. Hyperorthognathic	91.0 and above.

IV. Facial index (after Virchow) $\frac{100 \text{ facial height}}{\text{facial width,}}$ calculated from the linear distance of the facial breadth (No. 17) and the facial height (No. 19) (like the facial index of von Hölder).

Broad-face crania	to 90.0
Small-face crania	90.1 and over.

V. Upper facial index (after Virchow) $\frac{100 \text{ upper facial height}}{\text{facial width.}}$ calculated from the linear distance of the facial width (No. 17) and the upper facial height (No. 20).

Broad upper face crania, index	to 50.0
Narrow upper face crania, index	50.1 and over.

VI. Zygomatic face index (after Kollmann). $\frac{100 \text{ facial height.}}{\text{Zygomatic breadth.}}$ calculated from the greatest

distance between the zygomatic arches (No. 18) and facial height (No. 19), gives two grades:

Low, chamæprosopic, face crania .... to 90. 0
High, leptoprosopic, face crania .... 90. 1 and above.

VII. Zygomatic upper face-height index (after Kollmann). $\frac{100 \text{ upper face height.}}{\text{Zygomatic breadth.}}$

Chamæprosopic upper face with index .... **to 50. 0**
Leptoprosopic upper face with index........................... 50. 1 and above.

VIII. Orbital index. $\frac{100 \text{ orbital height.}}{\text{Orbital width.}}$

Chamækonchic.. **to 80. 0**
Mesokonchic... 80. 1—85
Hypsikonchic... 85. 1 and over.

IX. Nasal index. $\frac{100 \text{ width of nasal cavity.}}{\text{Nasal height.}}$

Leptorrhinic... to 47. 0
Mesorrhinic... 47. 1—51. 0
Platyrrhinic... 51. 1—58. 0
Hyperplatyrrhinic .... 58.1 and over.

X. Palate index (after Virchow). $\frac{100 \text{ palate breadth.}}{\text{Palate length.}}$

Leptostaphylin... to 80. 0
Mesostaphylin... 80. 1—85. 0
Brachystaphylin .... 85. 1 and over.

Appendix C.

There are to be distinguished the following periods: First period of childhood, second period of childhood, adult age, ripe age, senility. These indications are enough and are worth more than those of years of age because the anatomical and physiological phenomena which they demonstrate are more or less precocious according to individuals or according to race. In our race these periods correspond nearly to the following ages:

First period of childhood, from birth to the end of the sixth year:

Second period of childhood	7 to 14 years.
Youth	14 to 25 years.
Adult age	25 to 40 years.
Ripe age	40 to 60 years.
Senility	beyond 60 years.

We give these figures as a concession to custom and to make the succession of the periods more easily appreciated. But let us hasten to add that they are for the most part very uncertain. * * *

First period of childhood.—From birth to the eruption of the first great molars, called sixth year's teeth. * * *

Second period of childhood.—It commences at about the age of six years with the eruption of the first permanent molar, which marks the beginning of the second dentition; it ends about the age of thirteen or fourteen years, when the eruption of the four second permanent molars is completely achieved. * * *

Youth.—It commences when the eruption of the four second permanent molars is completely achieved—that is to say, when the crowns of these teeth are altogether on a level with those of the first molars; it is finished when on the one hand the wisdom teeth are come out, and when on the other hand the basilar suture is completely closed.

Adult age, ripe age, and senility.—Onward from the end of the third period the distinction of ages is much more doubtful. It is based upon the observation of two phenomena which are gradual and very irregular in their chronology. * * *

Physiologically one is generally contented to divide all the time which passes from the end of youth to death into two periods: The period of gradual change, called indifferently virility, adult age, or ripe age, and the period of decadence, called senility. In craniology the first of these periods should be divided in two ages: Adult age, comprised between the end of youth and the beginning of the ossification of the sutures, and the period of gradual change from then on to senility. The craniological distinction between adult age and ripe age is generally easy since it rests upon the anatomical observation of the study of the sutures. * * *

Senility of the skull is recognized by the following characters:

First. The sutures are mostly in an advanced or complete state of ossification: some of them at least are entirely effaced and may have left not even a vestige. The others, with the exception of the squamous suture, which sometimes remains open until a very advanced age, are more or less ossified. * * *

------

* Instructions craniologiques et craniometriques par P. Broca, Paris, 1875, pp. 128 et seq.

Second. The wearing away of the teeth which are yet in place is very pronounced. * * * The *alveolar point* mounts almost up to the level of the nasal spine; the mandible reduces itself to its *basilar portion;* the height of the symphysis of the chin is found reduced more than one-half, and finally the *angle of the jaw* becomes very obtuse. * * *

Third. The bones of the cranial vault of old persons sometimes are subject to an interstitial resorption of the spongy tissue; the two compact tables of the bone become fused in one compact and semitransparent plate, and from this result the undulating depressions characteristic of *senile atrophy,* which are the certain signs of an advanced old age. The most ordinary seat of these senile atrophies is the zone of the parietal comprised between the sagittal suture and the superior temporal line of that bone.

PLATE I.

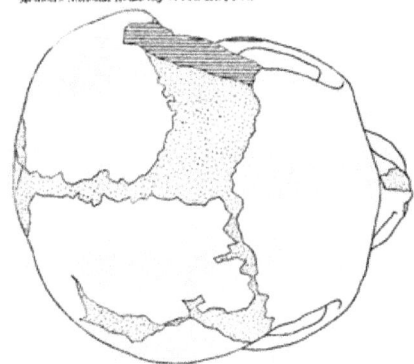

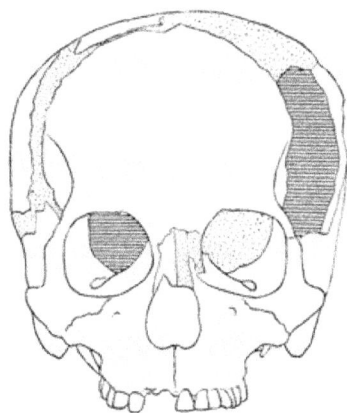

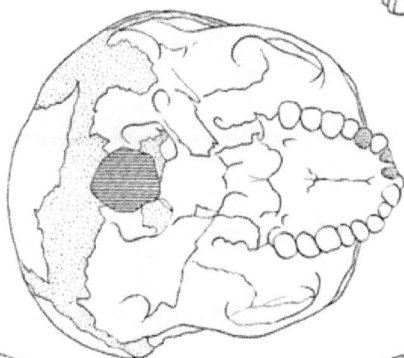

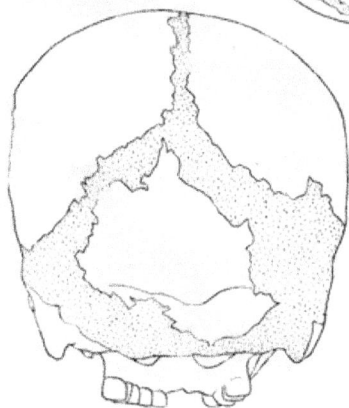

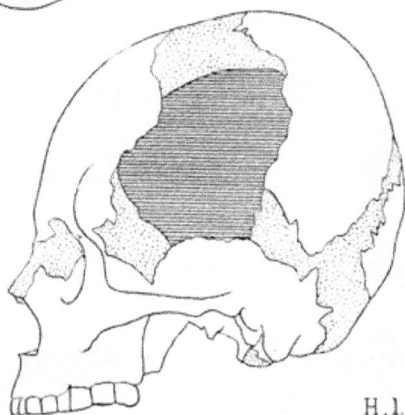

H. 1.

PLATE II.

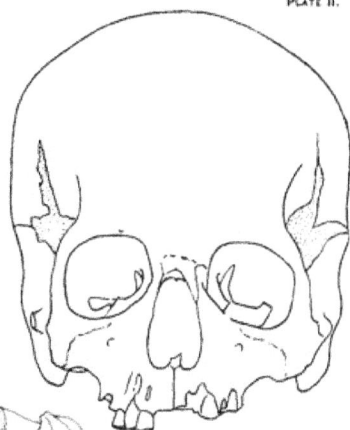

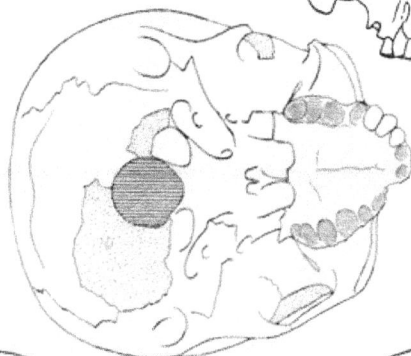

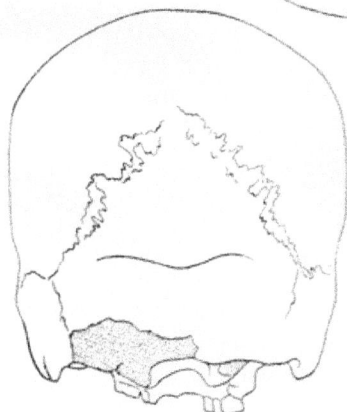

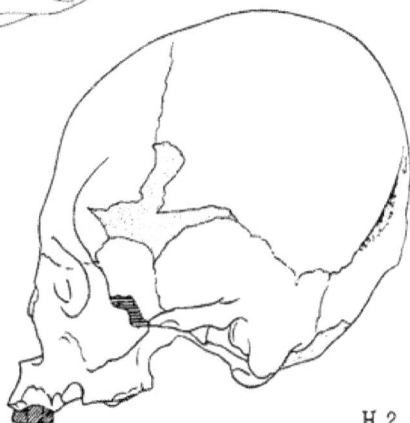

H. 2.

PLATE III.

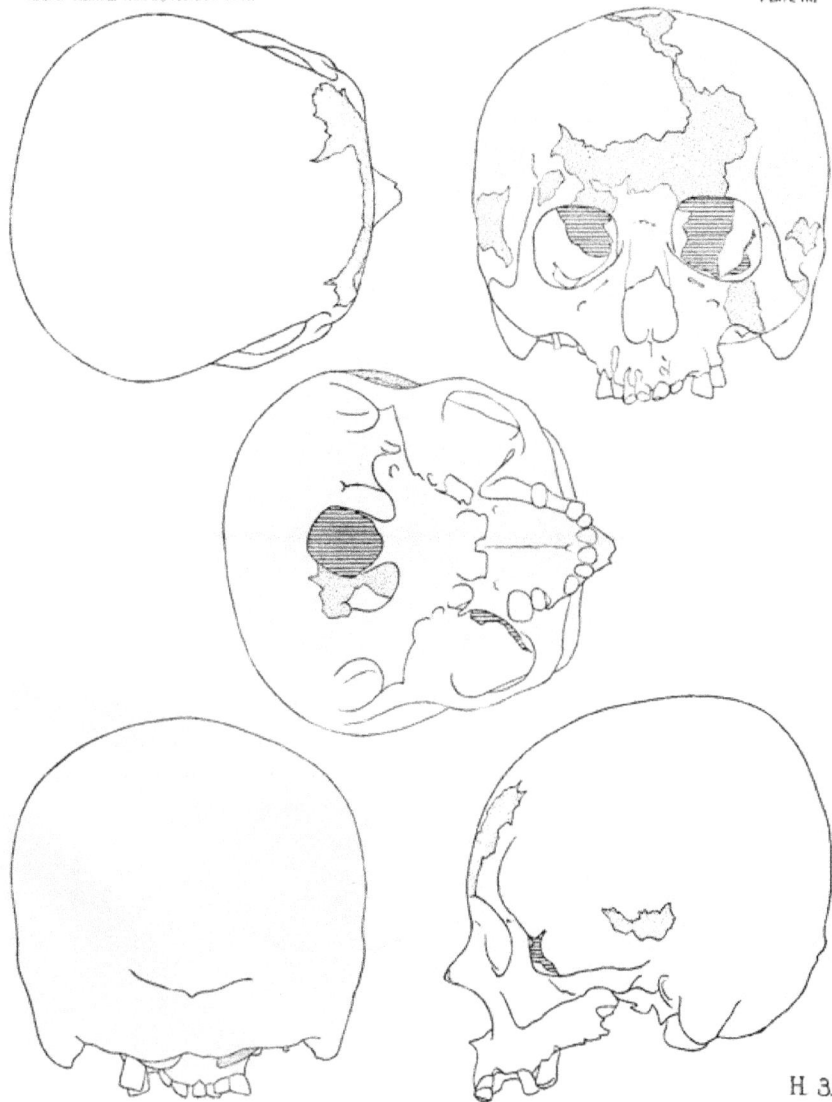

H 3.

PLATE IV.

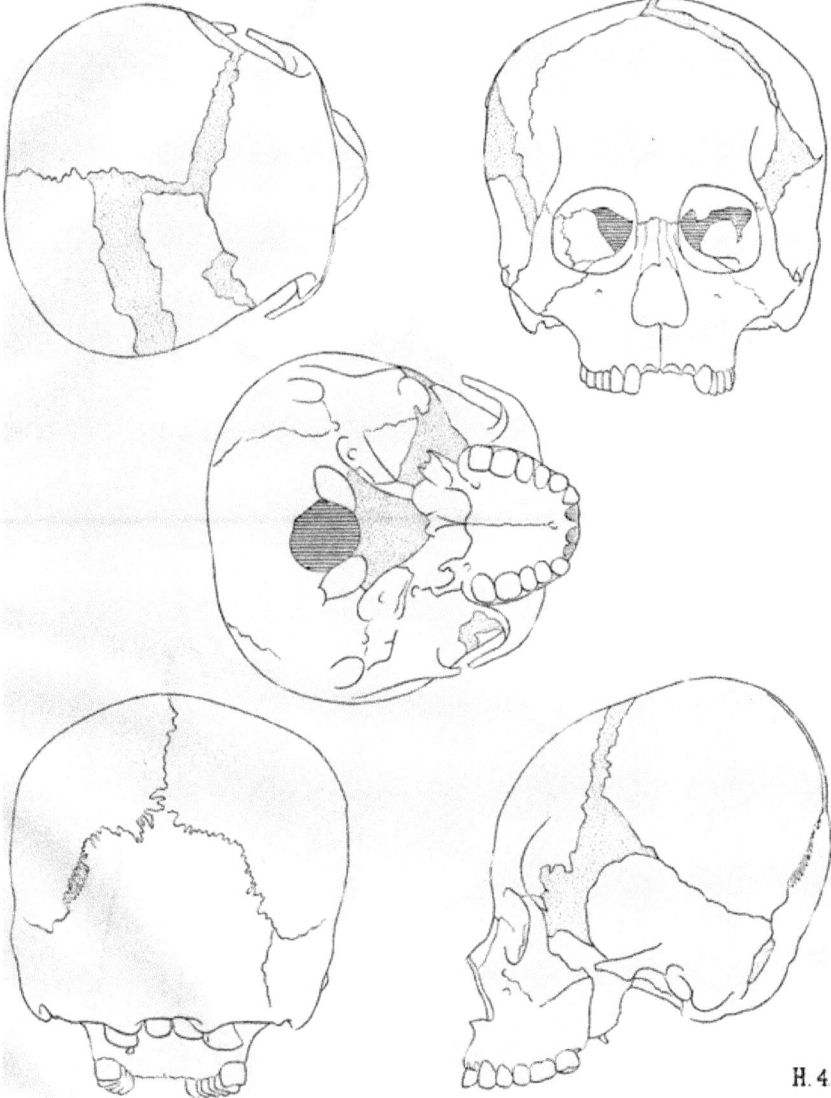

H. 4

PLATE V.

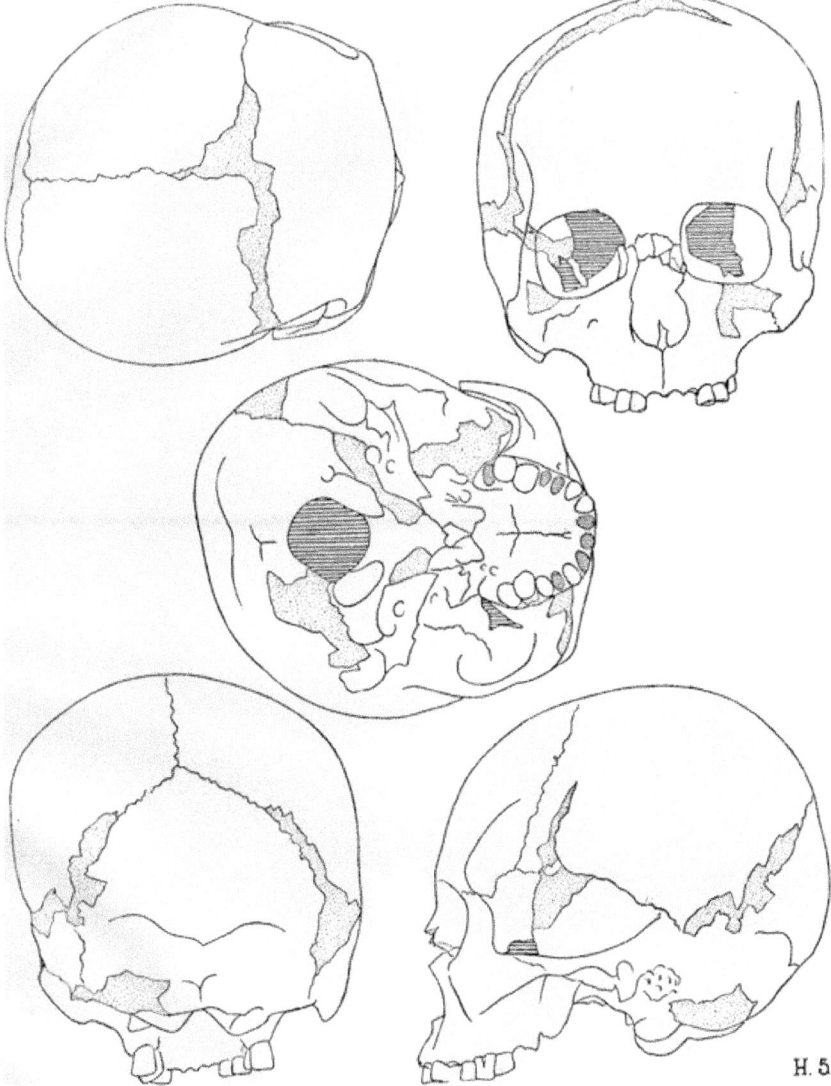

H. 5

PLATE VI.

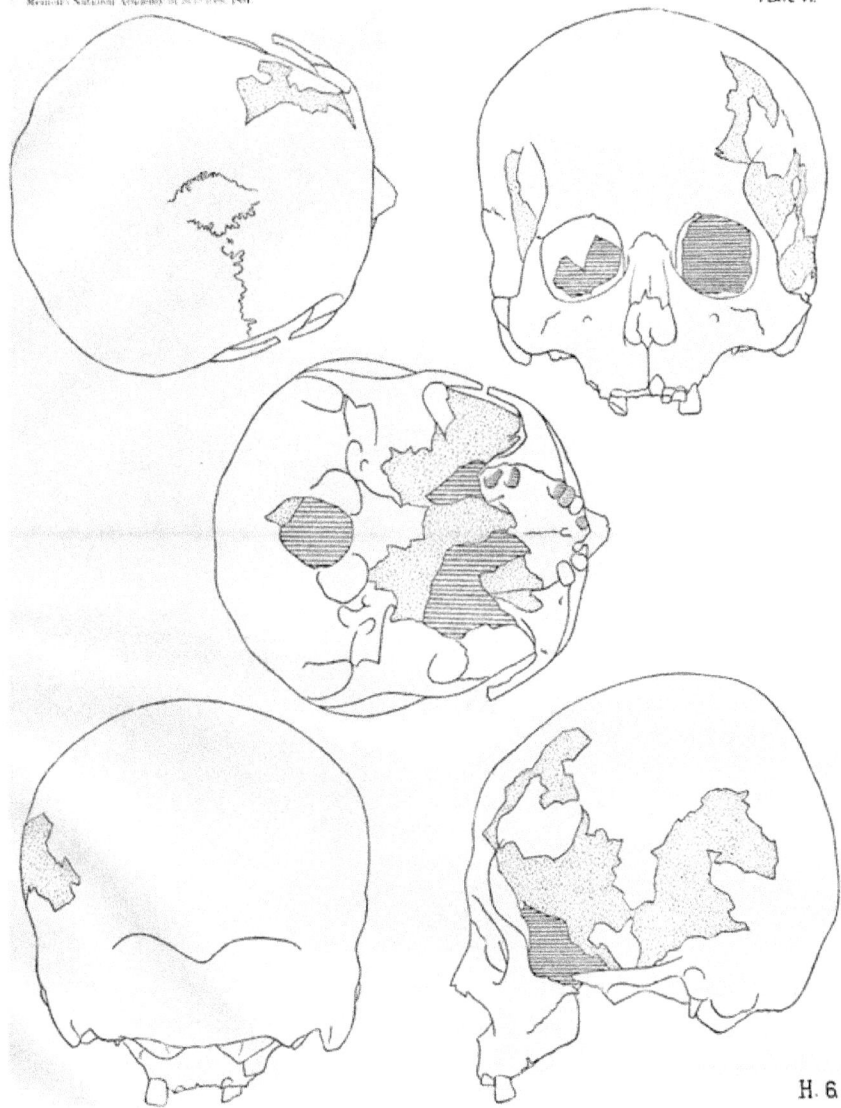

H. 6

PLATE VII.

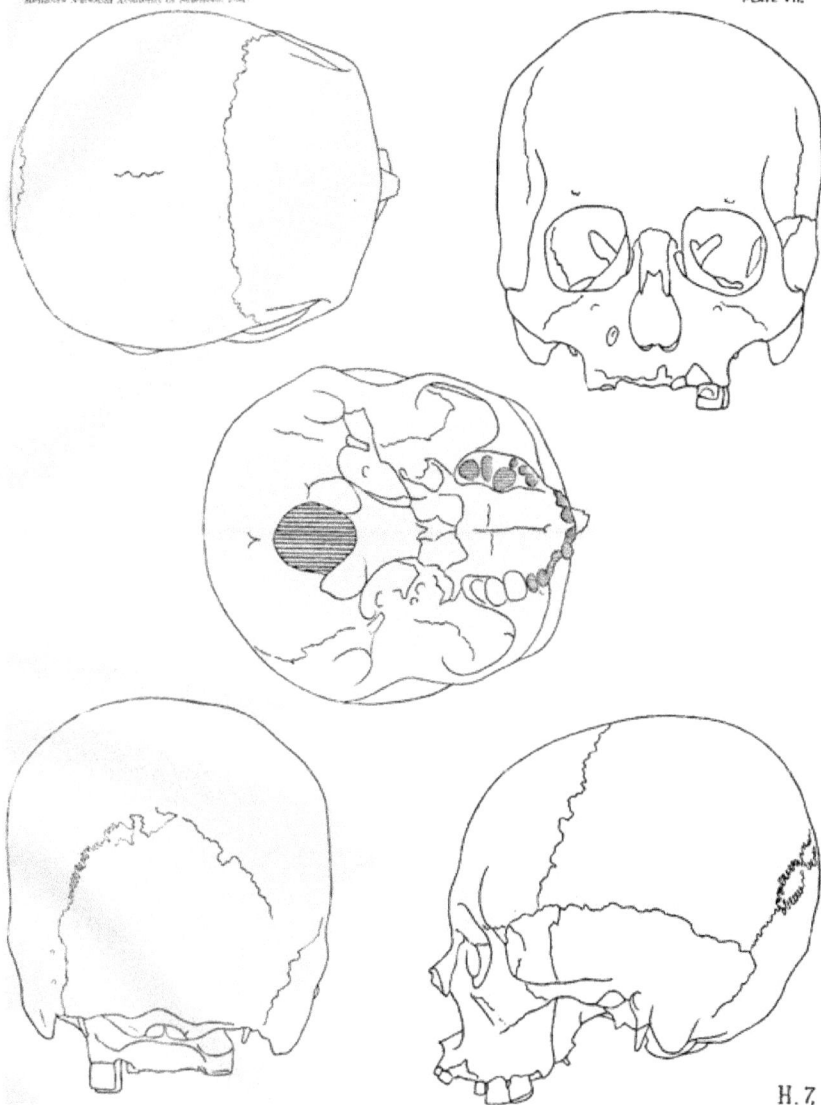

H. 7.

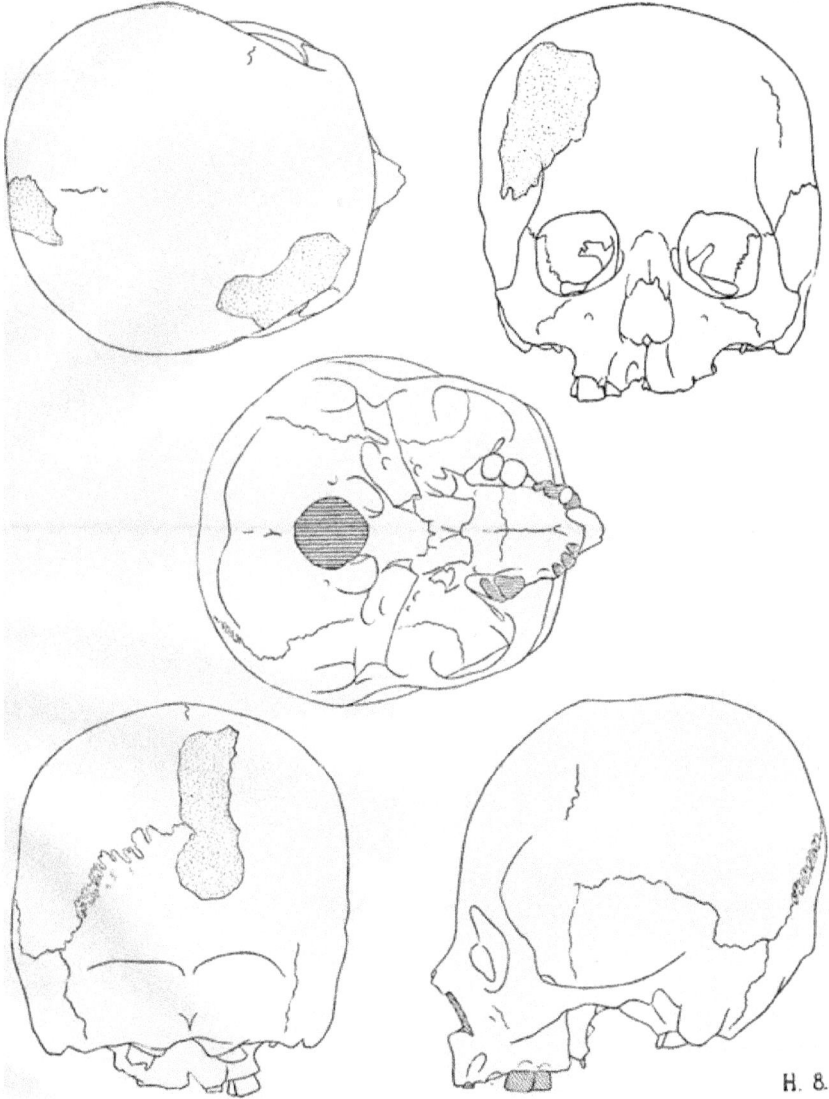

PLATE VIII

H. 8.

PLATE IX.

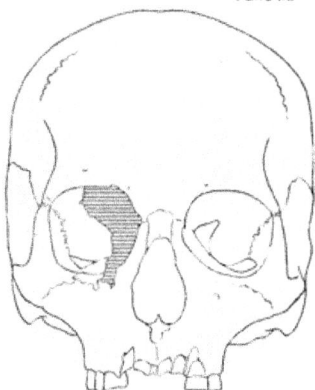

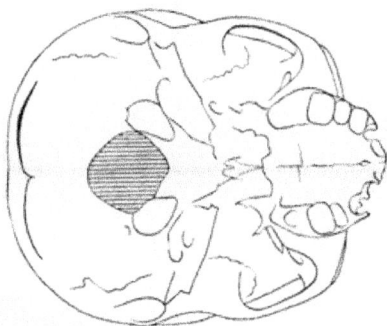

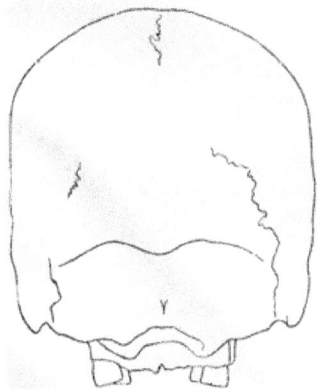

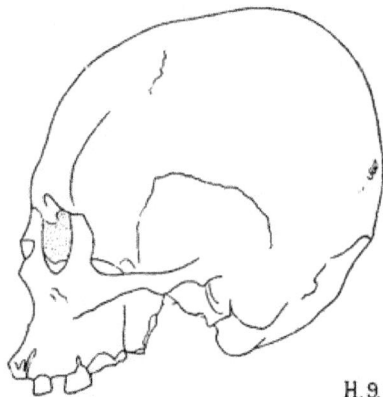

H.9.

PLATE X.

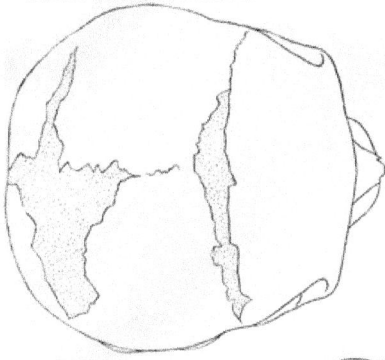

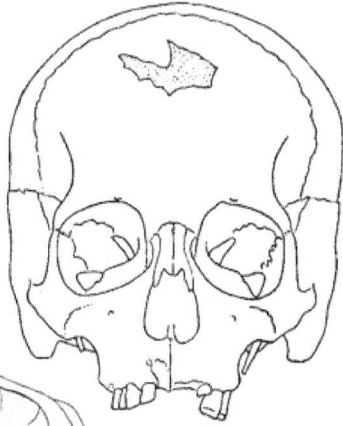

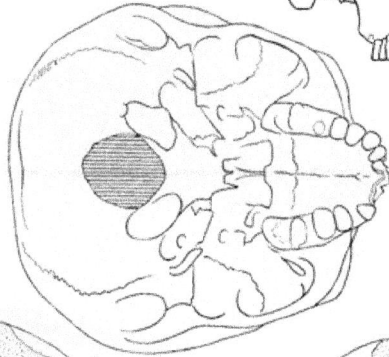

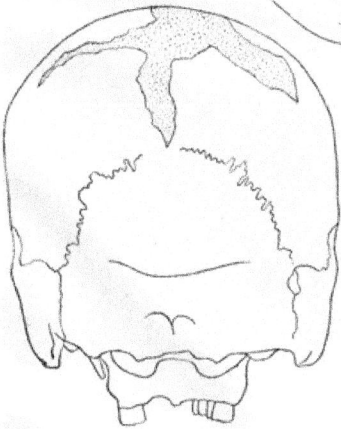

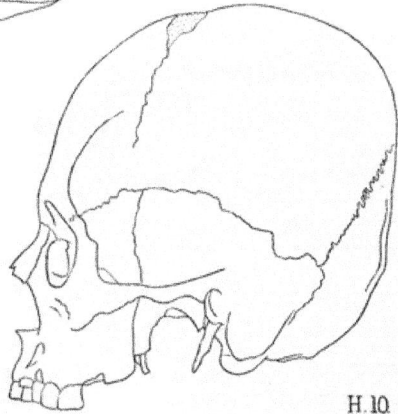

H. 10.

PLATE XI.

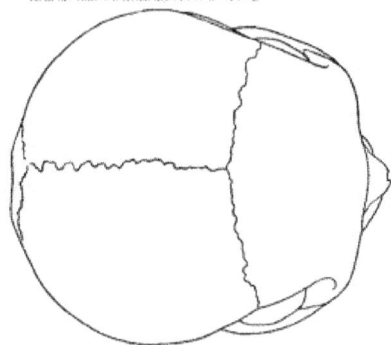

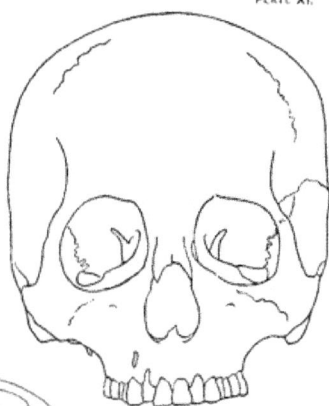

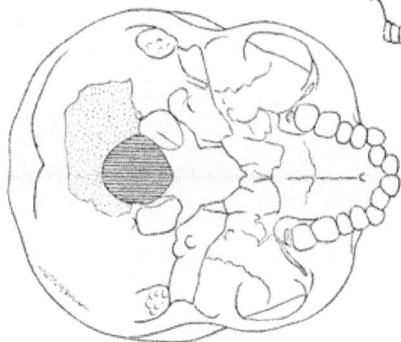

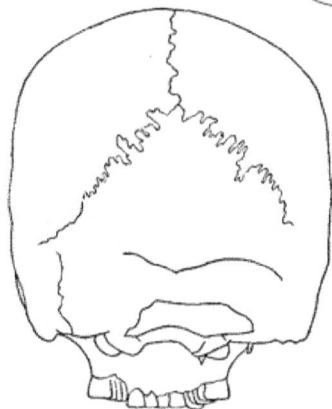

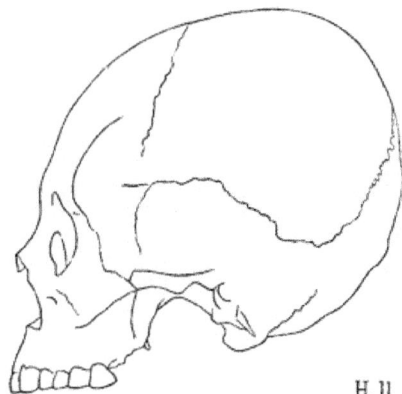

H. II

PLATE XII.

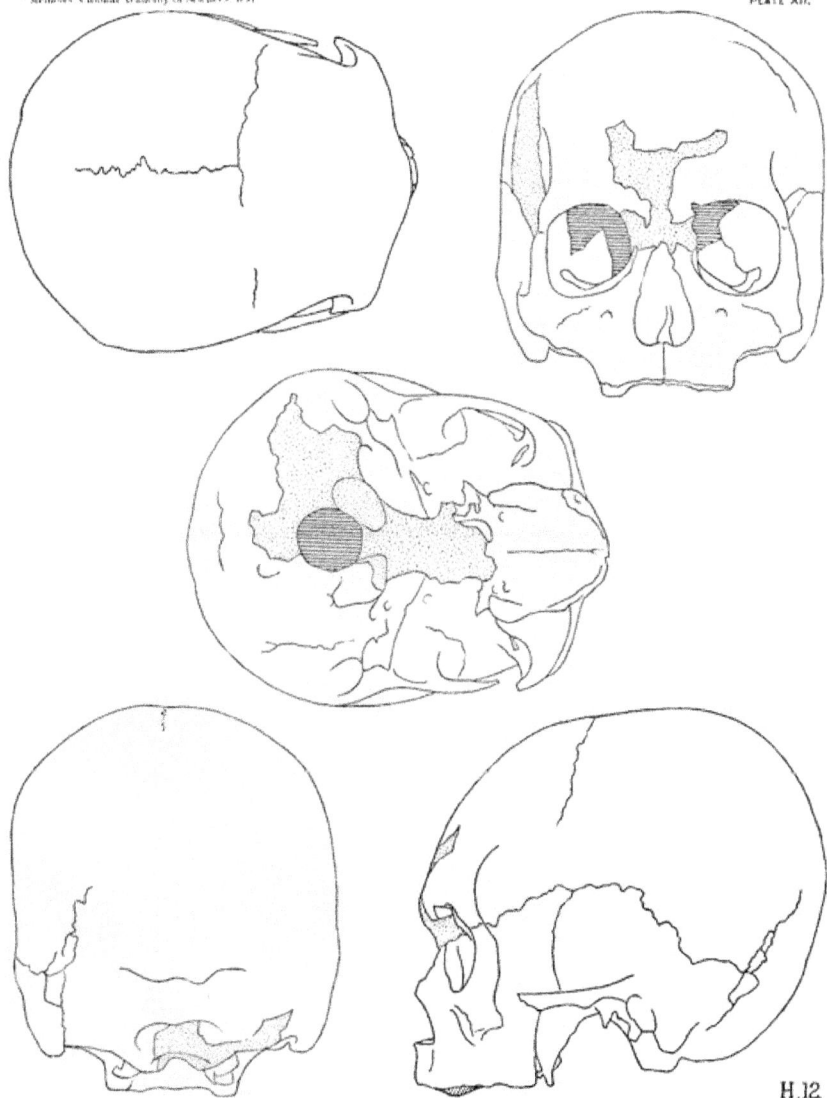

H.12.

PLATE XII.

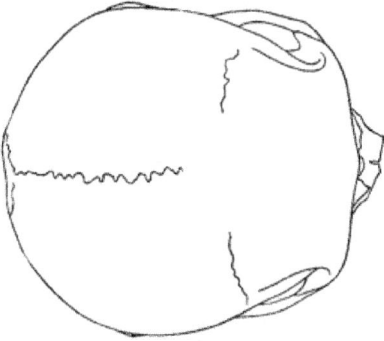

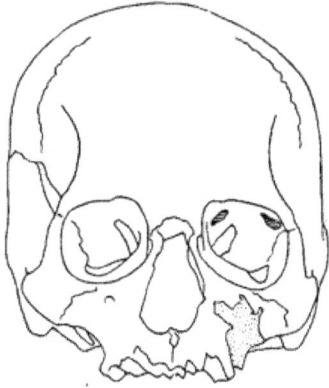

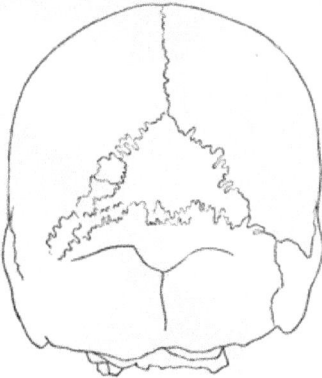

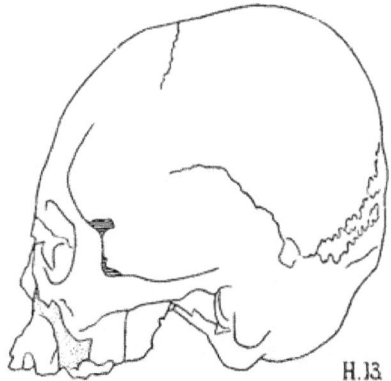

H.13.

PLATE XIV.

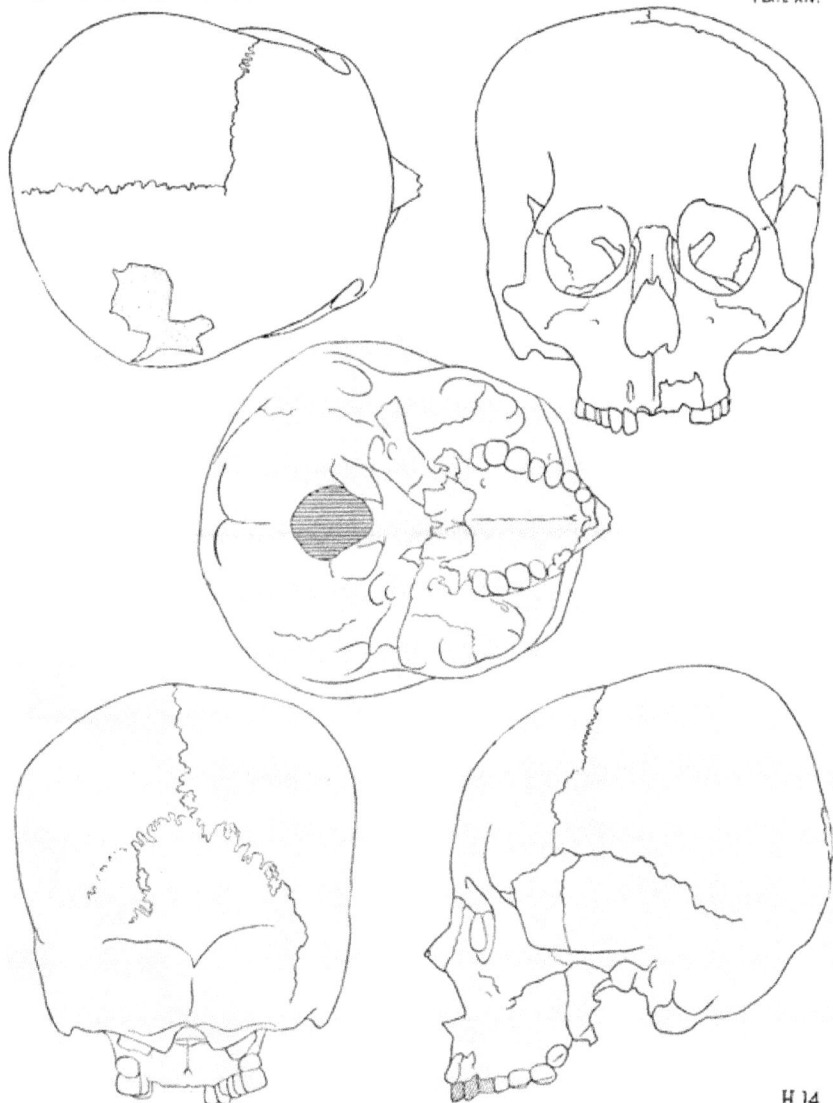

H. 14.

PLATE XV.

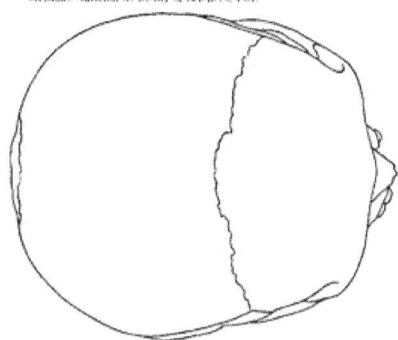

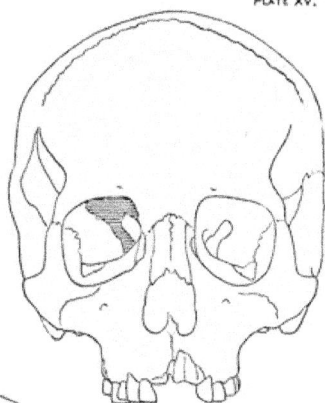

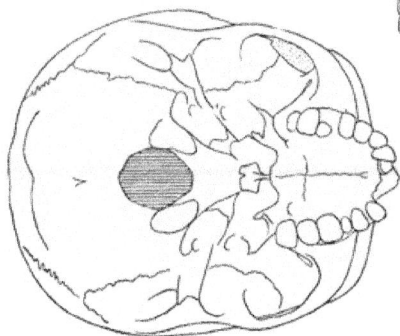

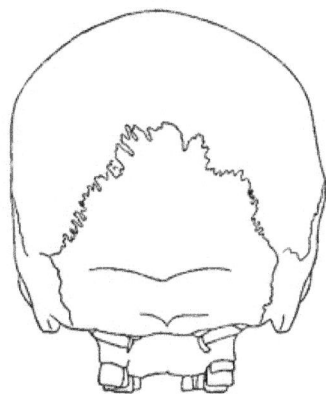

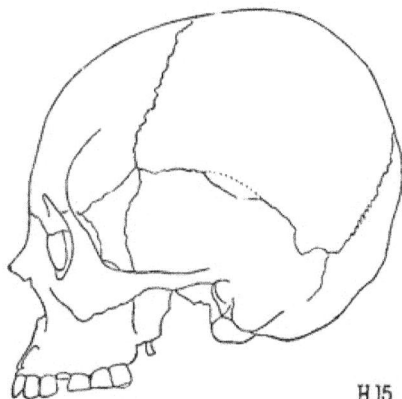

H.15.

PLATE XVI.

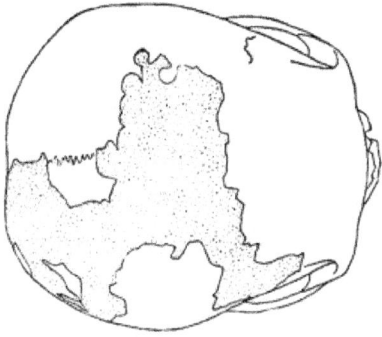

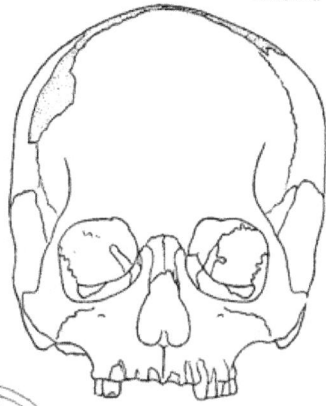

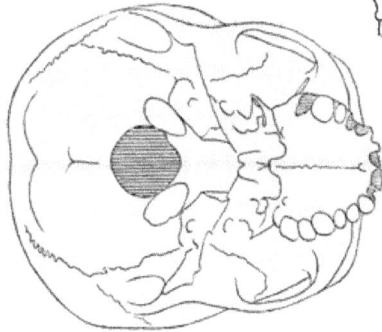

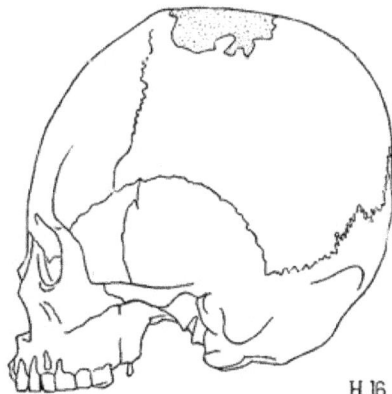

H.16

PLATE XVII.

H.17.

PLATE XVIII.

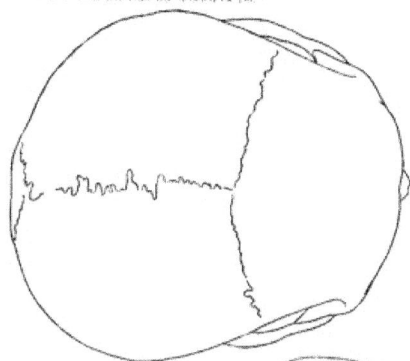
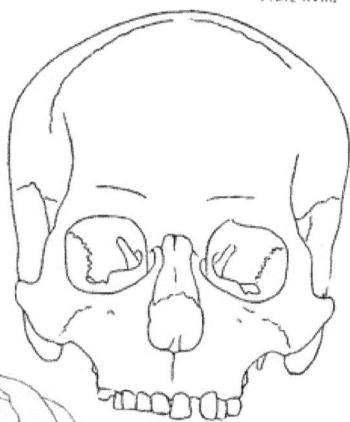
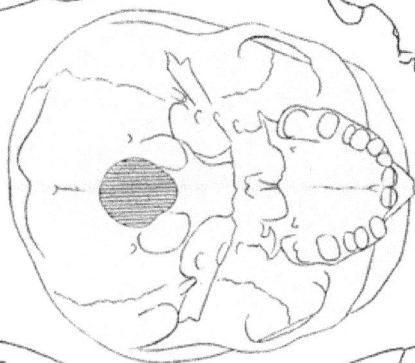
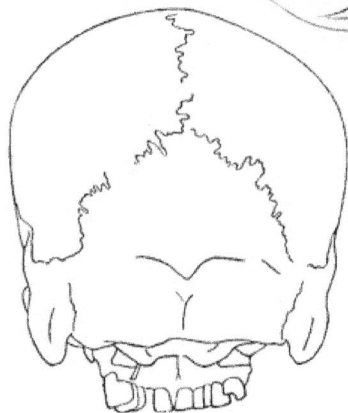
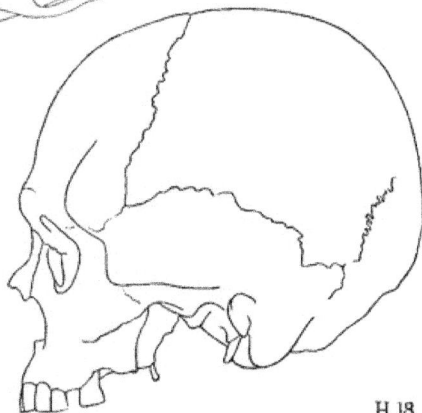

H 18.

PLATE XIX.

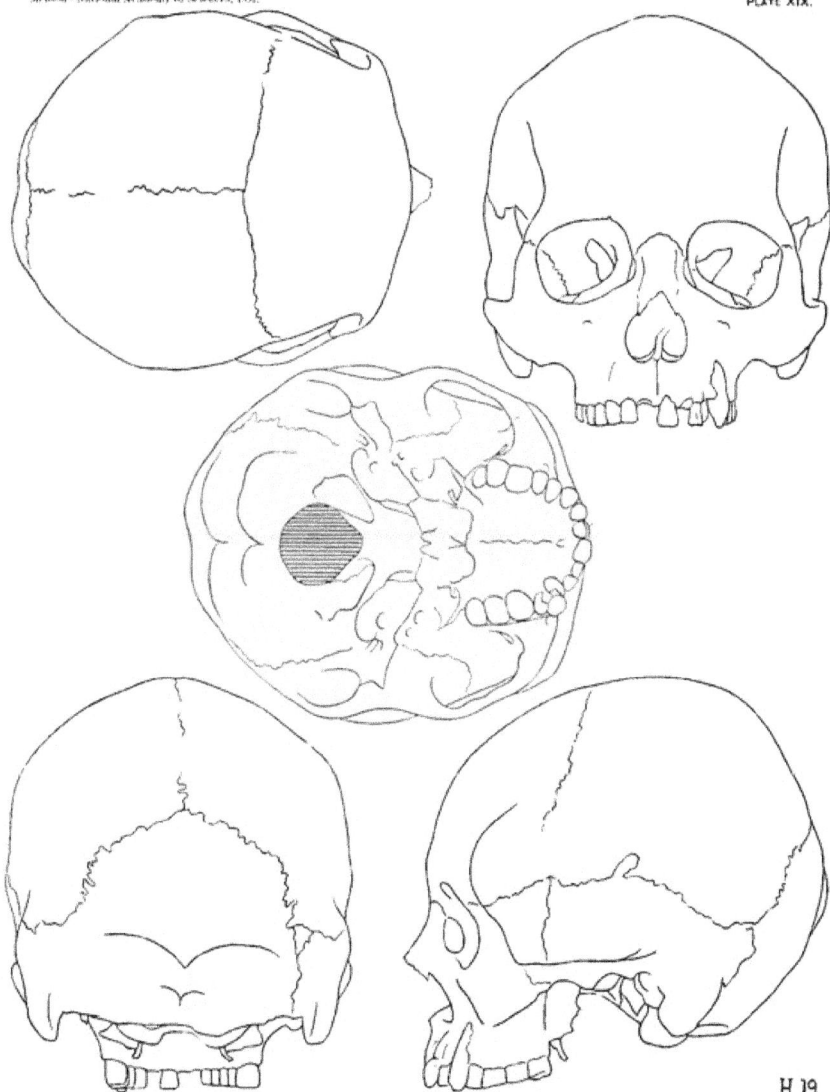

H.19.

PLATE XX.

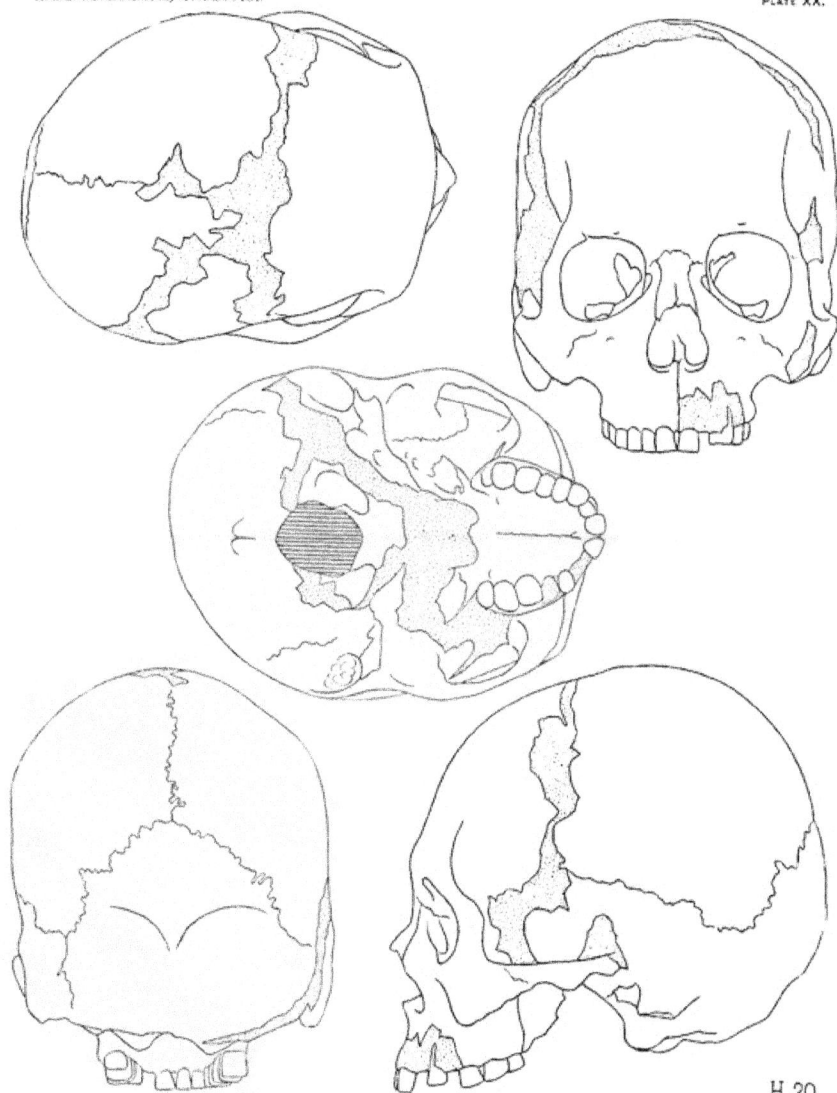

H. 20.

PLATE XXI.

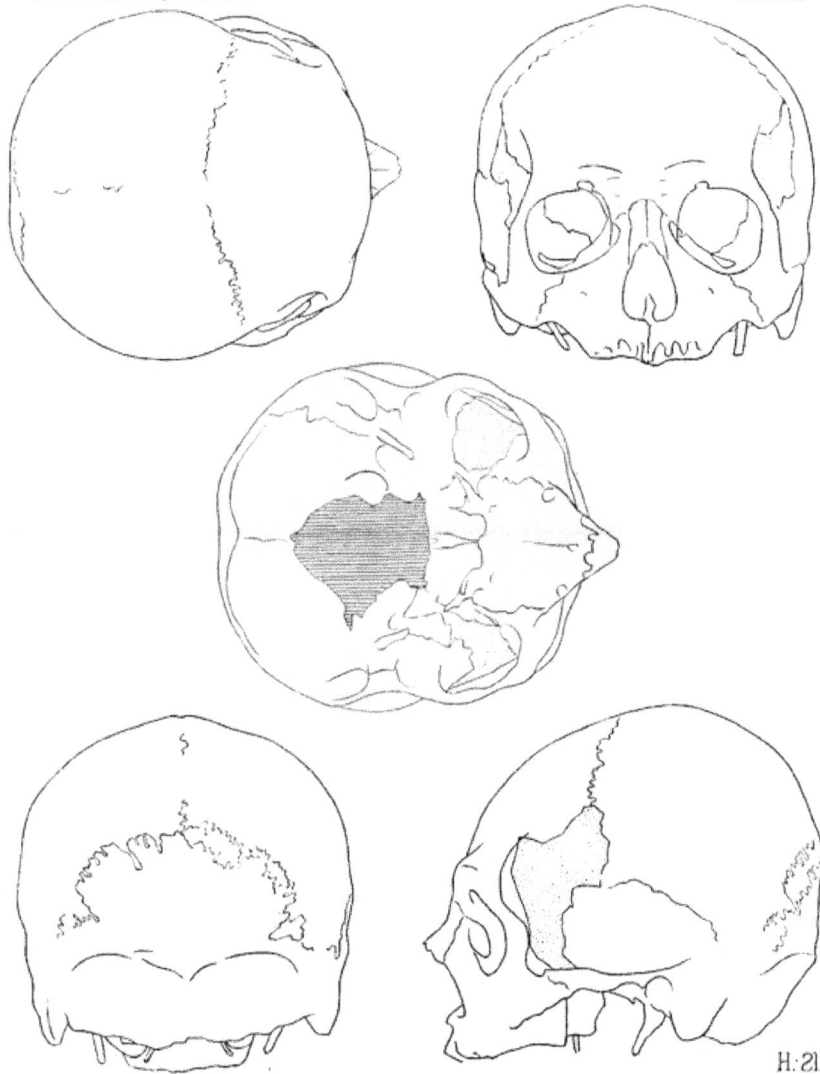

H: 21.

PLATE XXII.

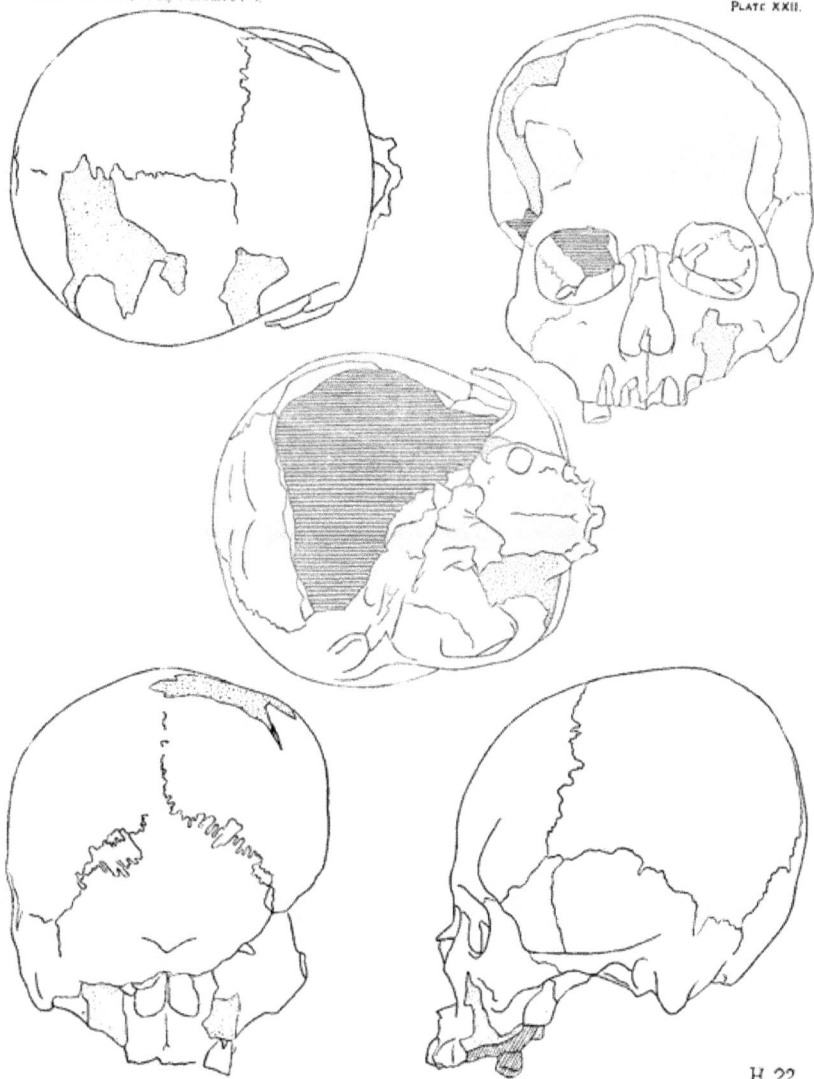

H. 22

PLATE XXIII.

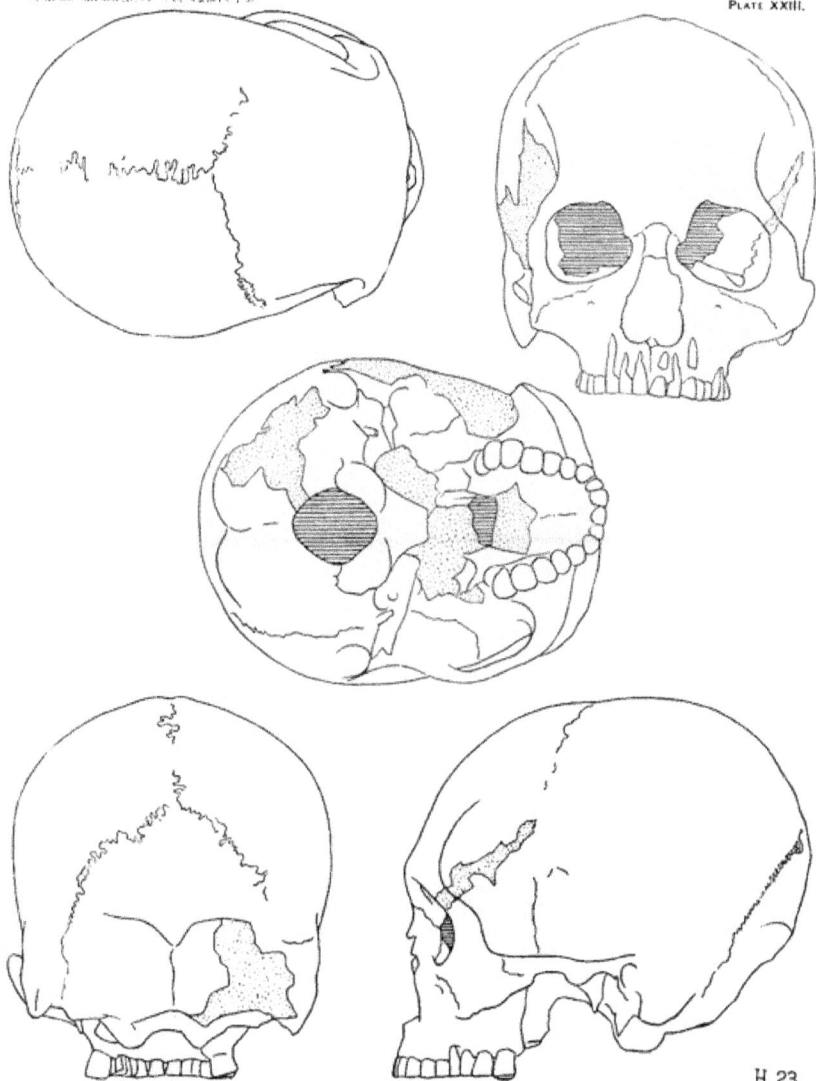

H. 23

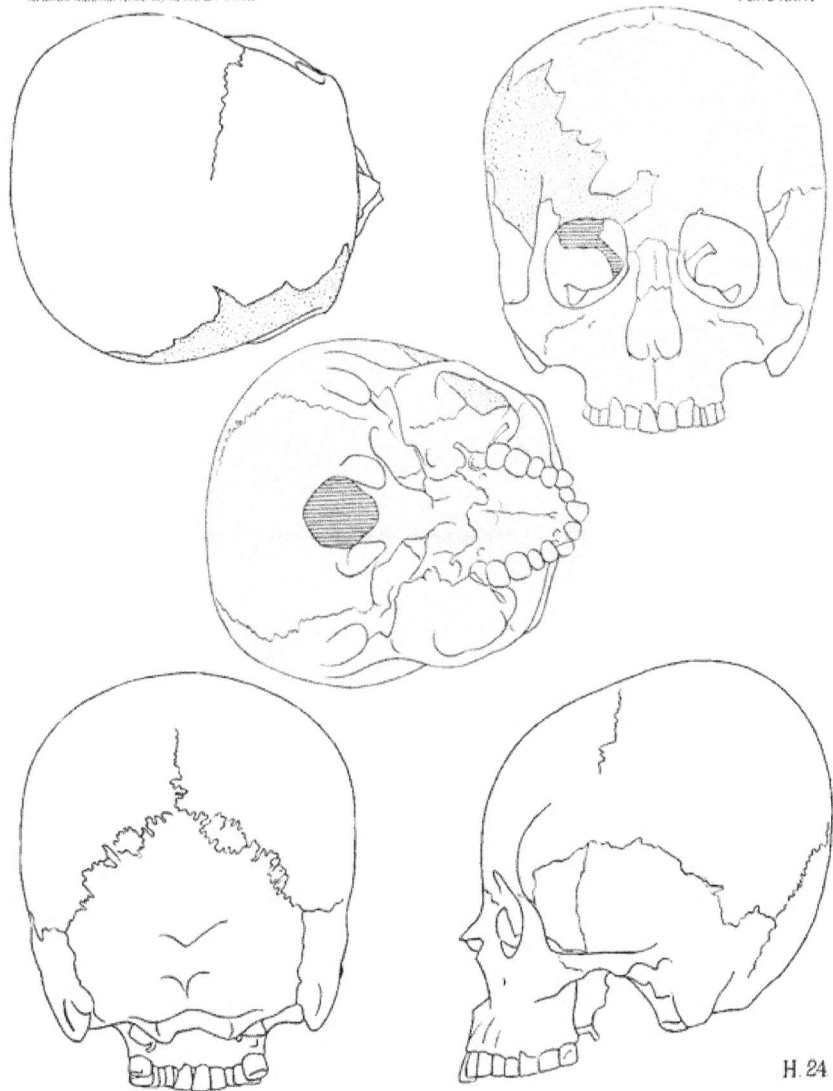

PLATE XXIV.

H. 24

PLATE XXV.

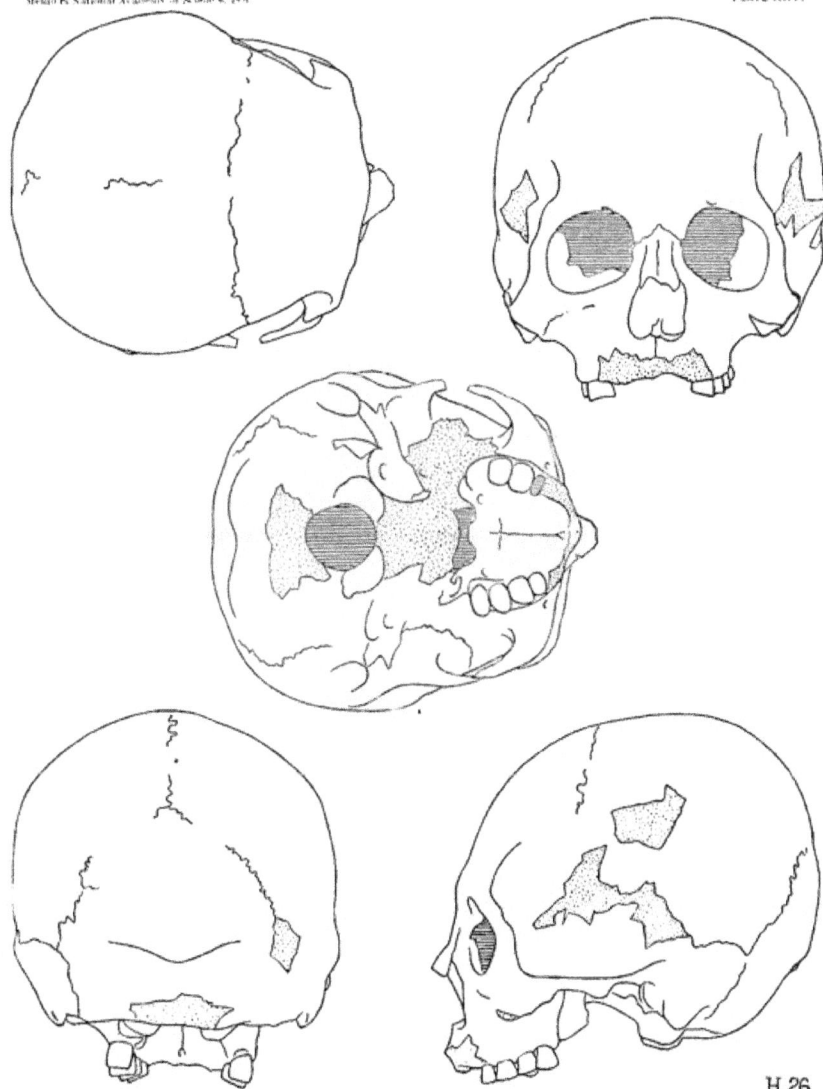

H. 26.

PLATE XXVI.

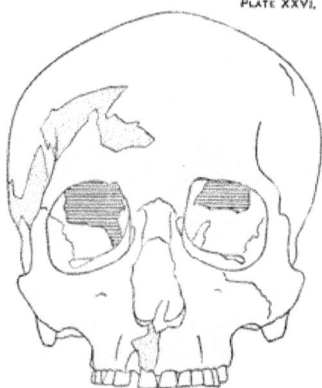

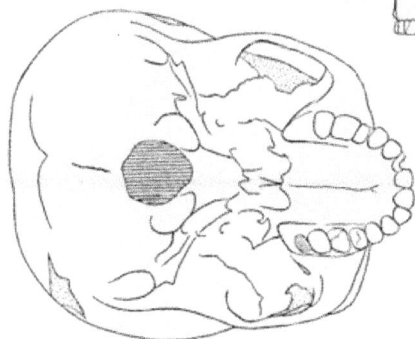

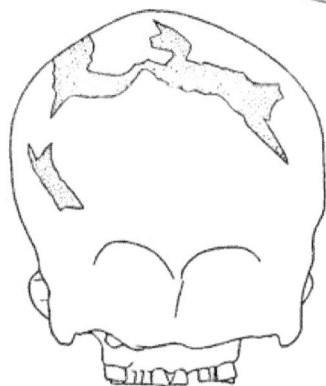

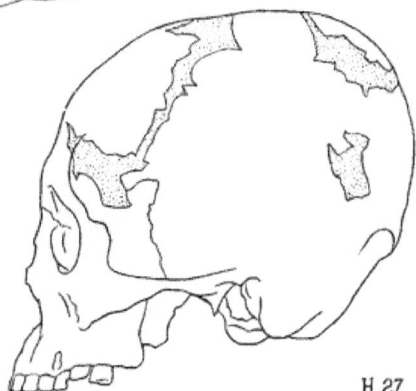

H. 27.

PLATE XXVII.

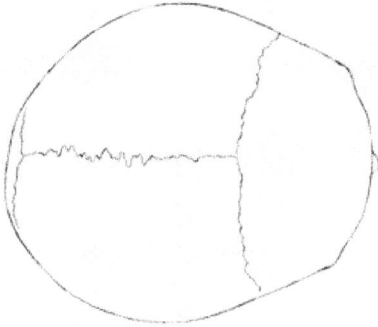

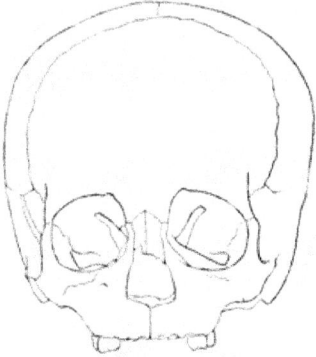

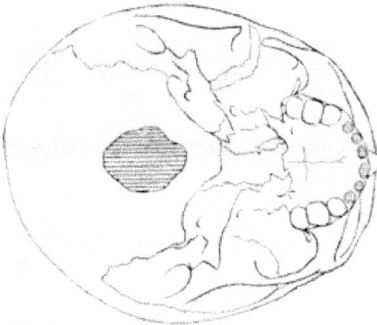

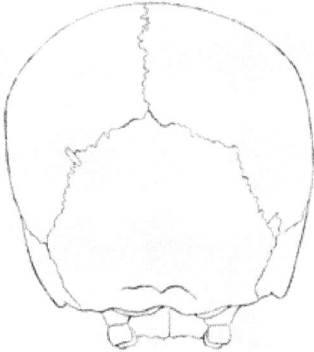

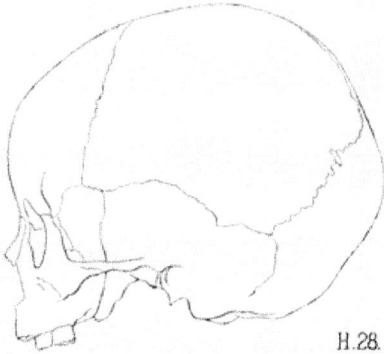

H.28.

PLATE XXVIII.

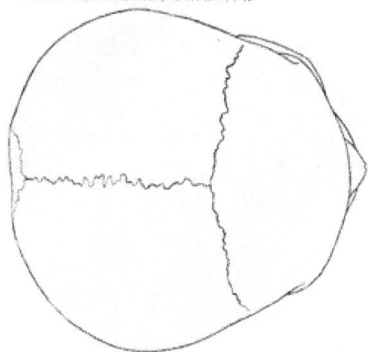

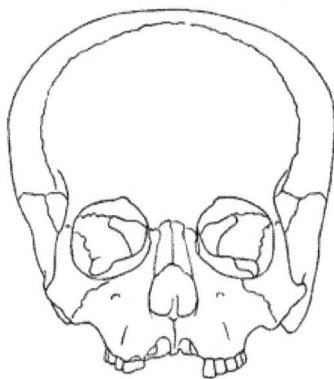

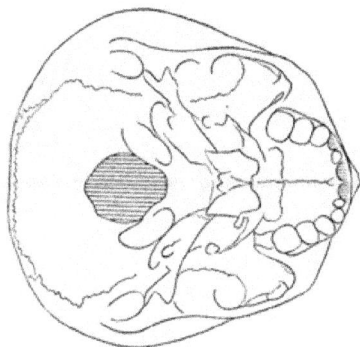

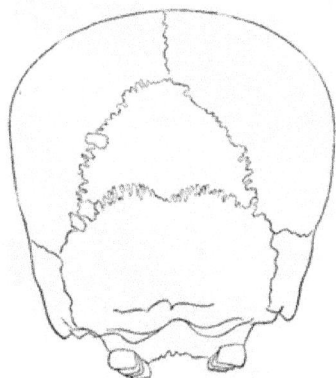

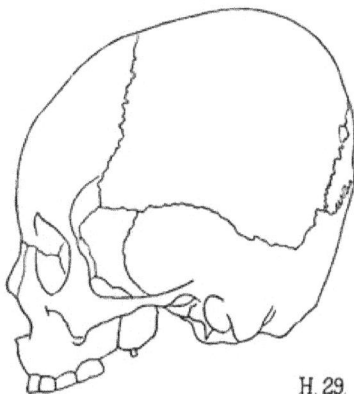

H. 29.

PLATE XXIX.

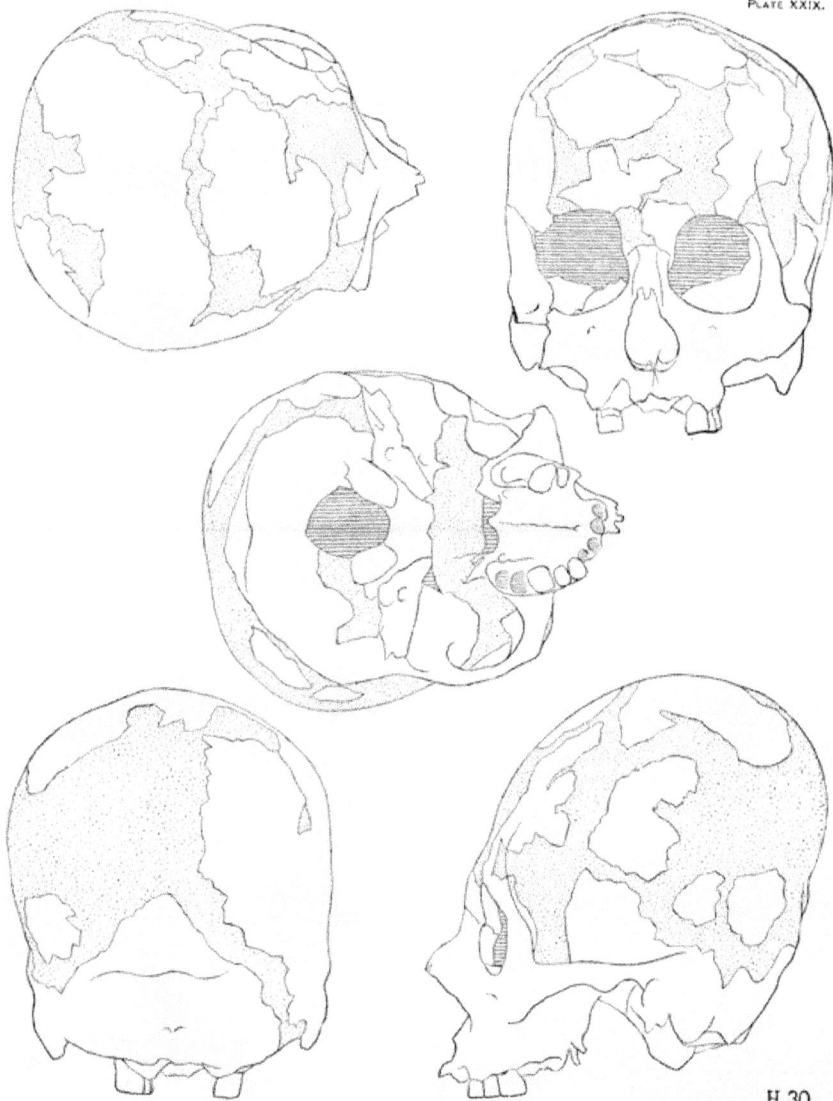

H. 30.

PLATE XXX

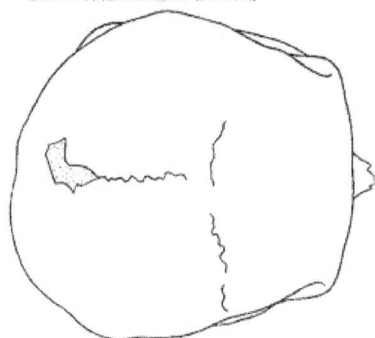
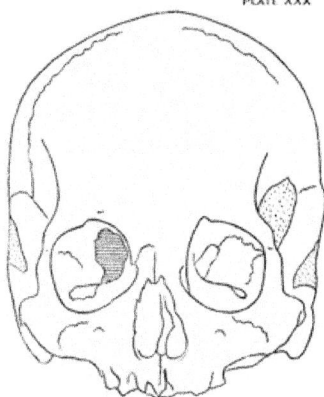
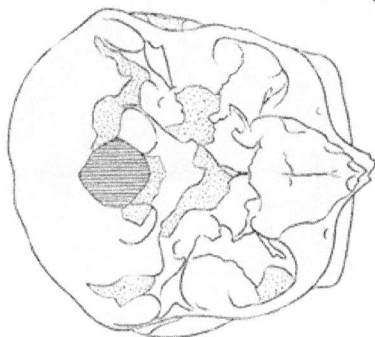
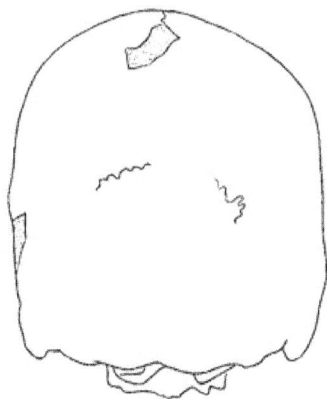
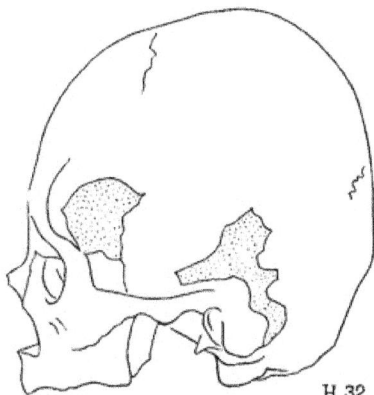

H. 32.

PLATE XXXI.

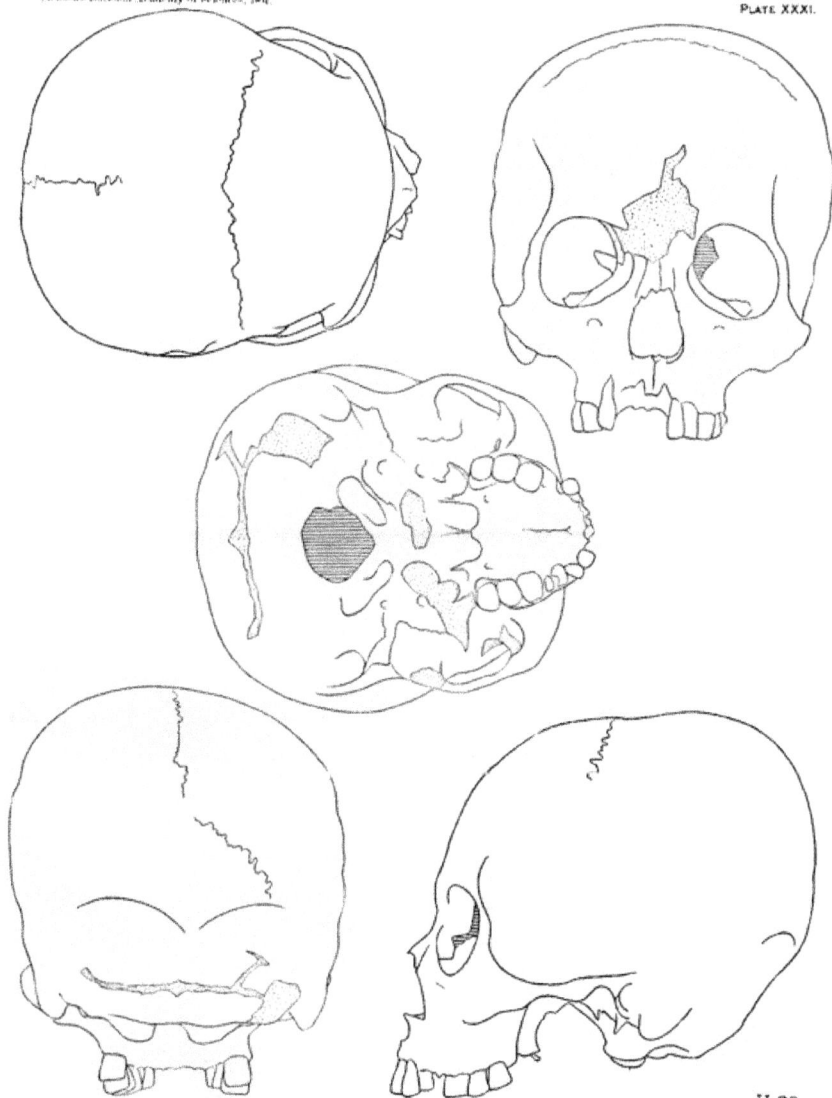

H.33

PLATE XXXII.

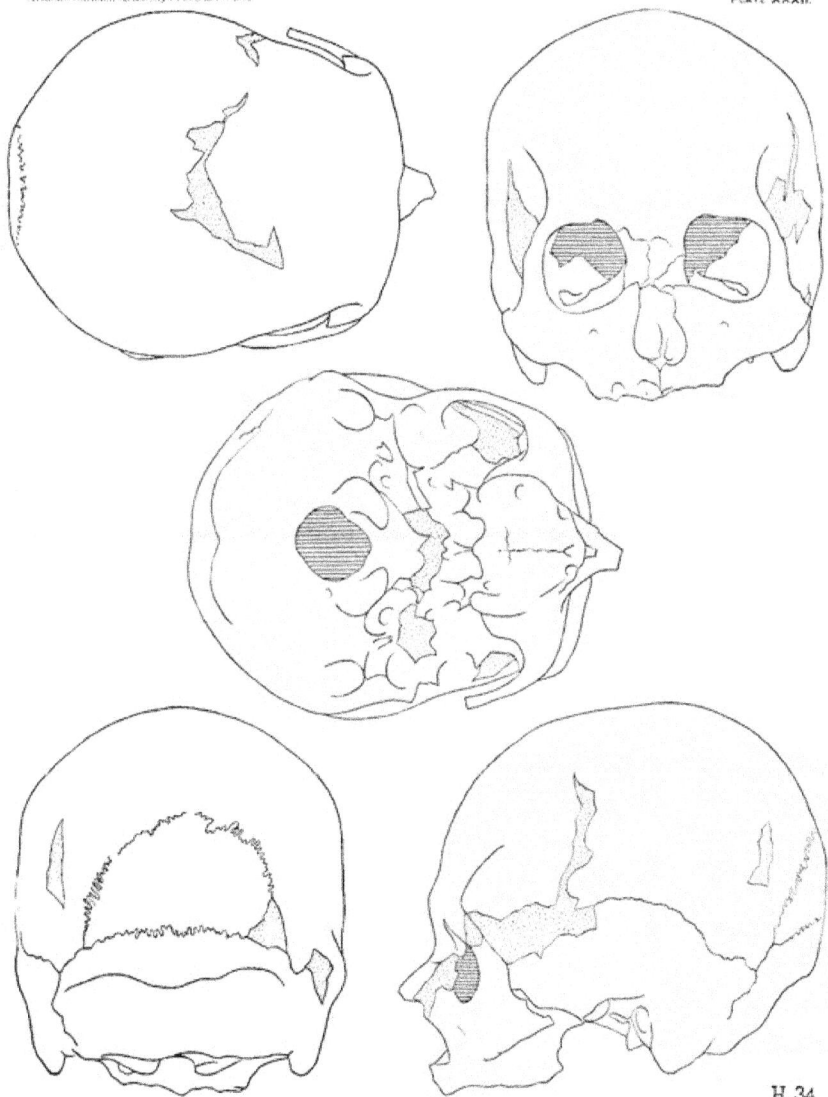

H.34

PLATE XXXIII.

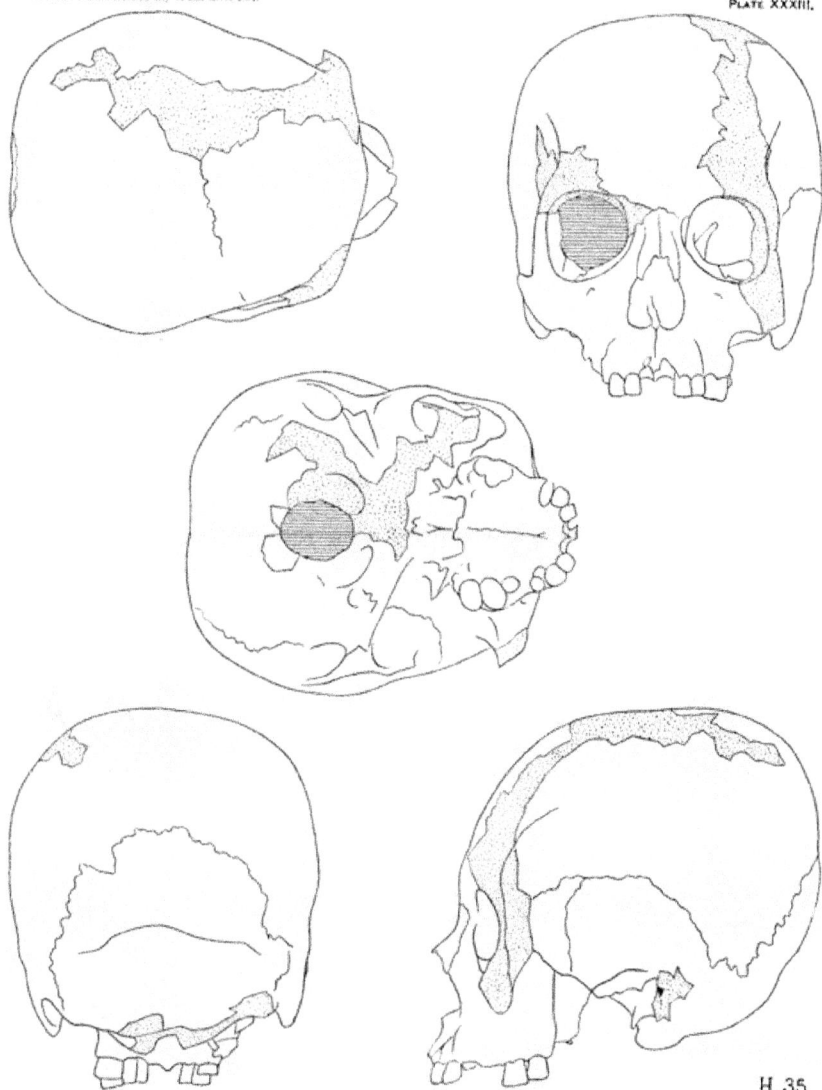

H. 35.

PLATE XXXIV.

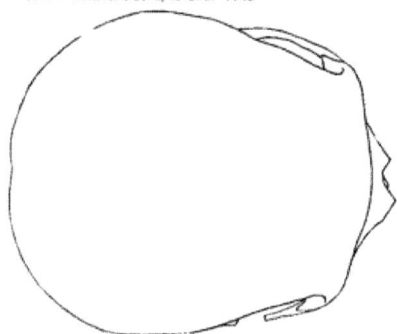

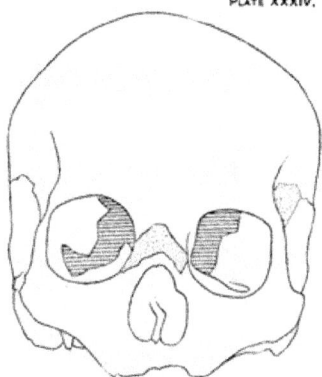

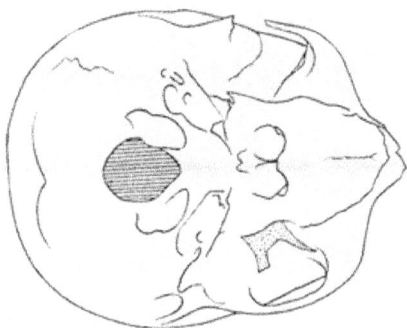

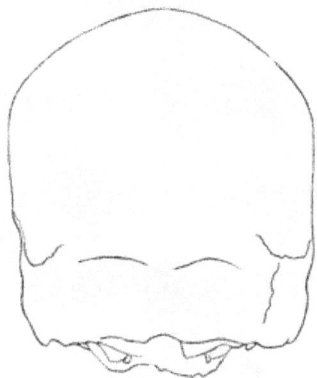

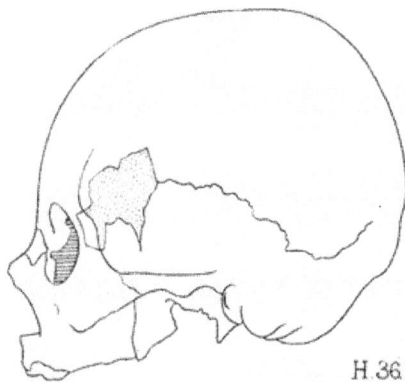

H.36

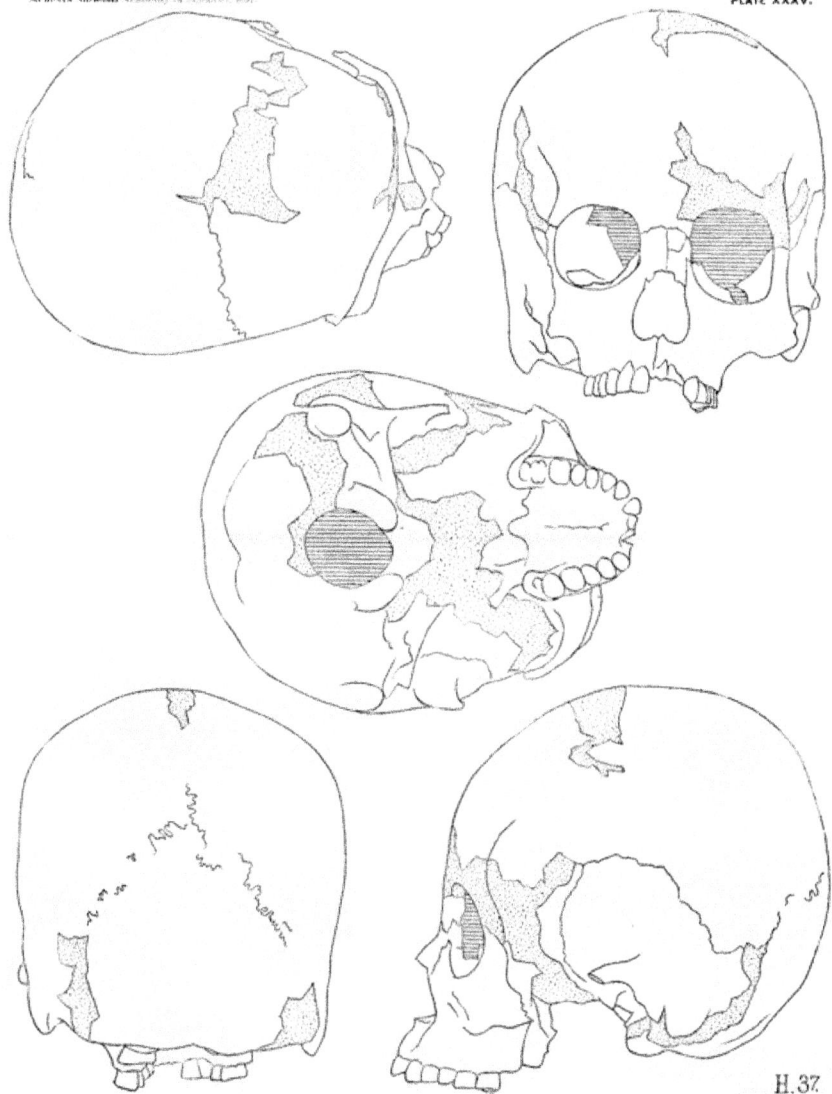

PLATE XXXV.

H.37.

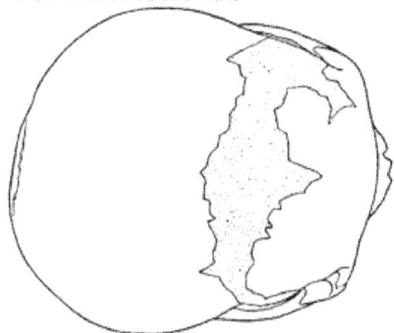

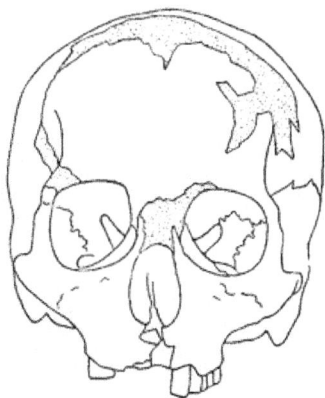

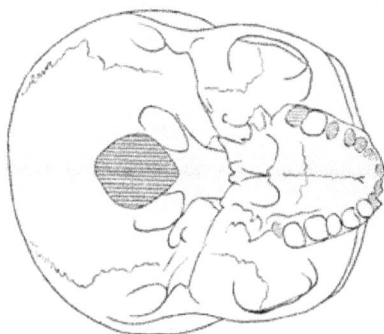

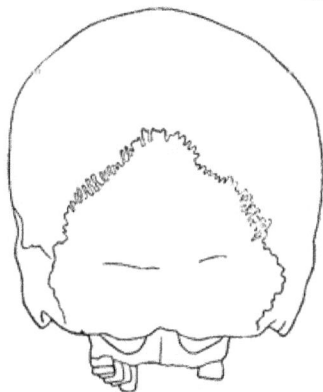

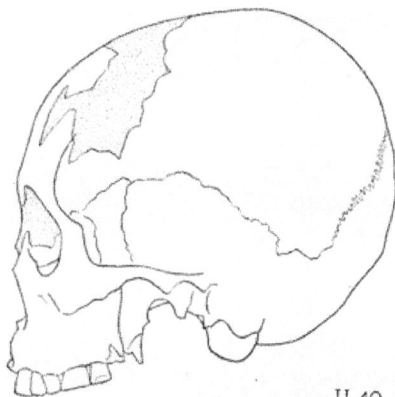

Plate XXXVI.

H.40.

PLATE XXXVII.

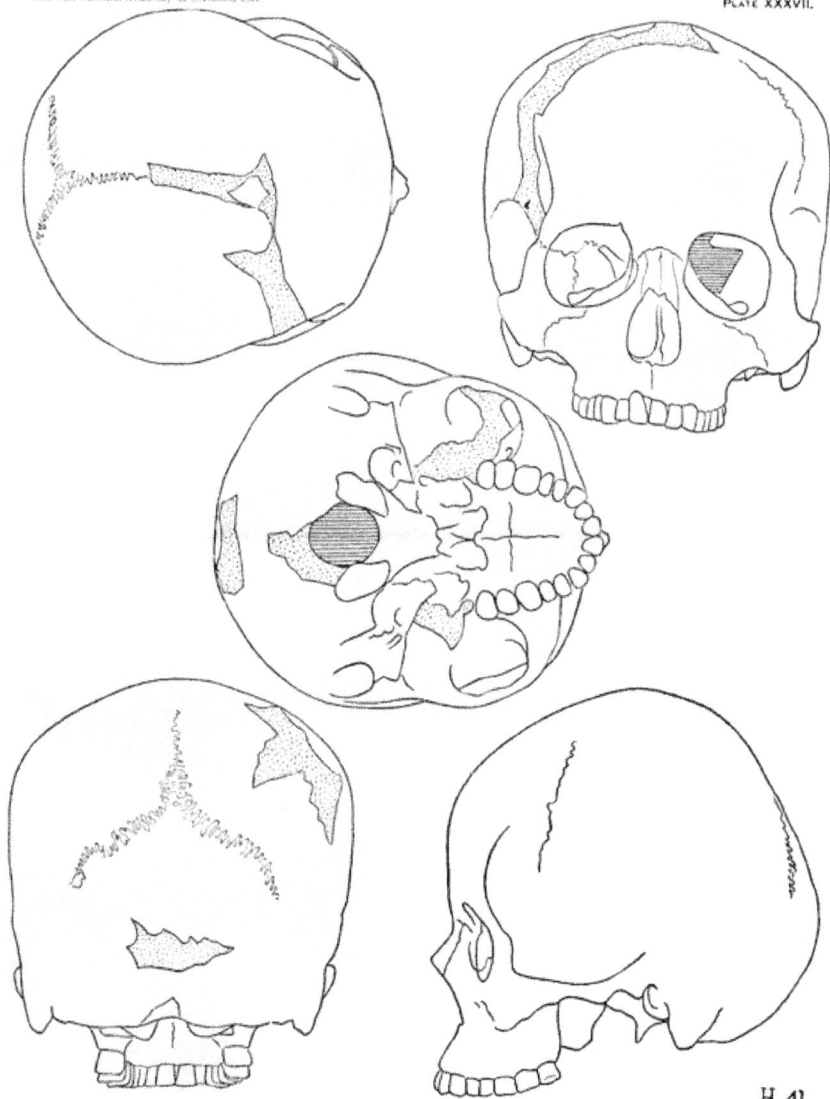

H. 41.

PLATE XXXVIII.

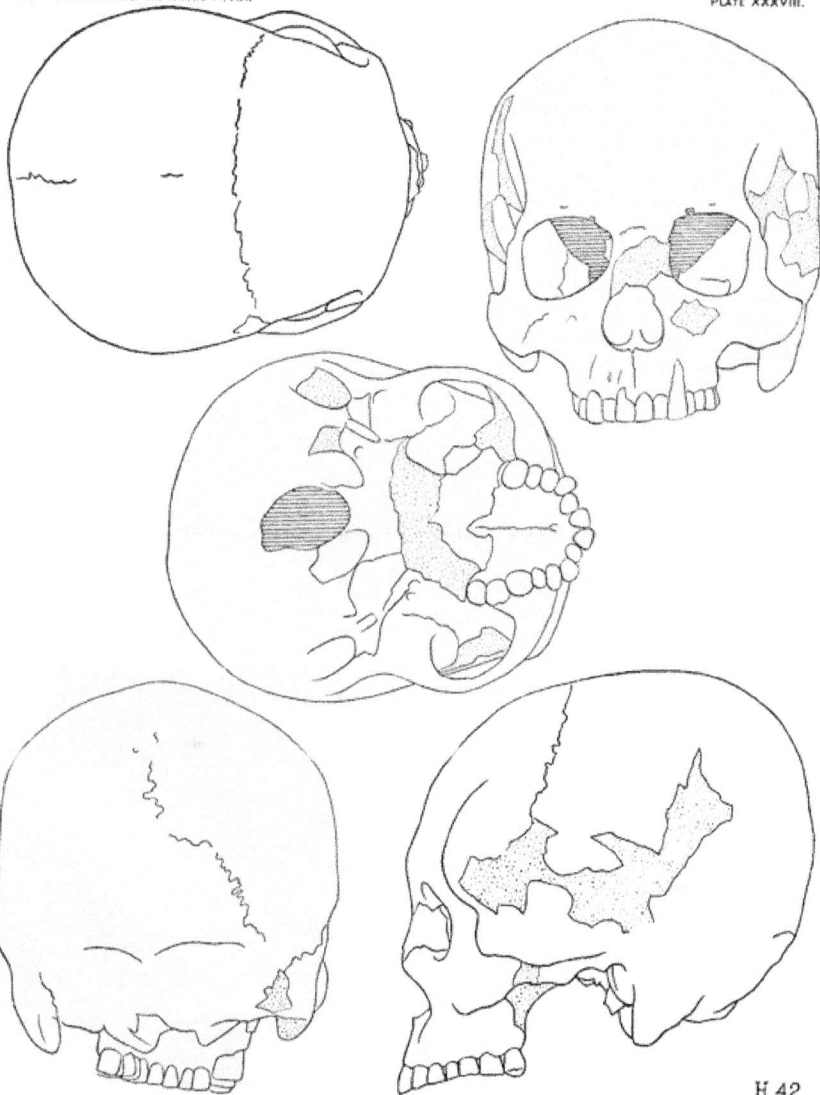

H.42.

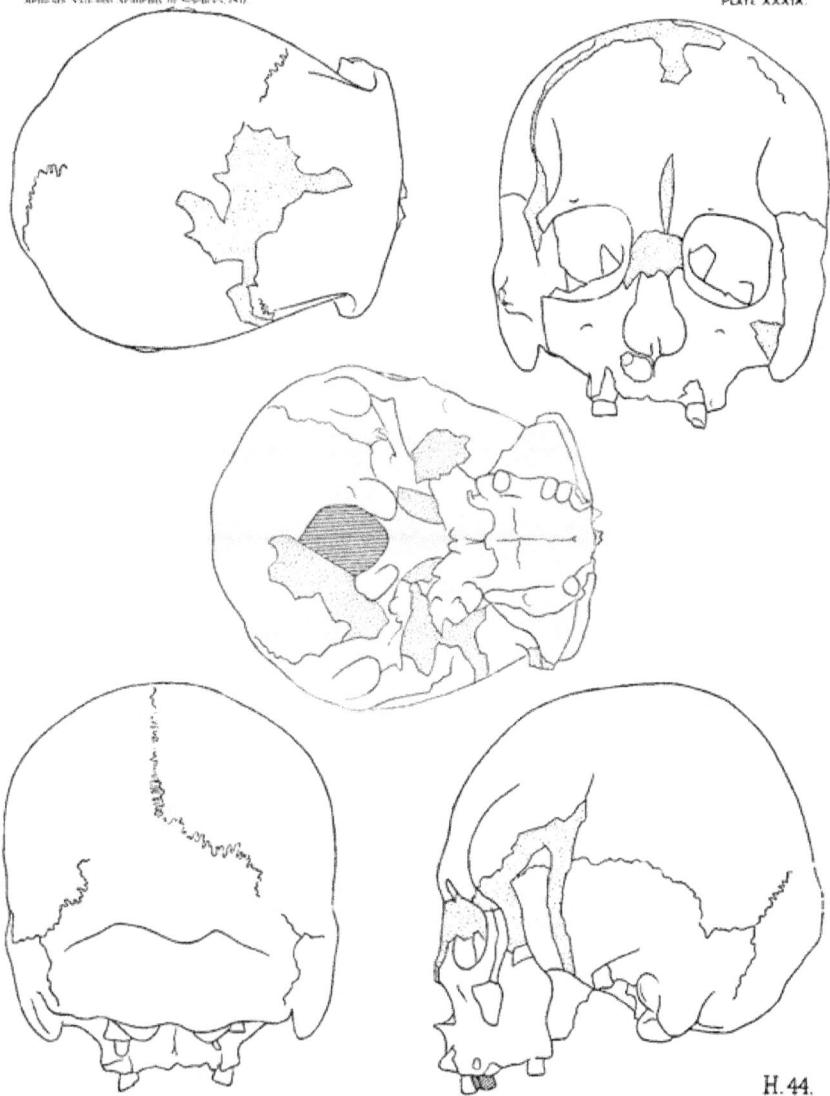

H. 44.

PLATE XL.

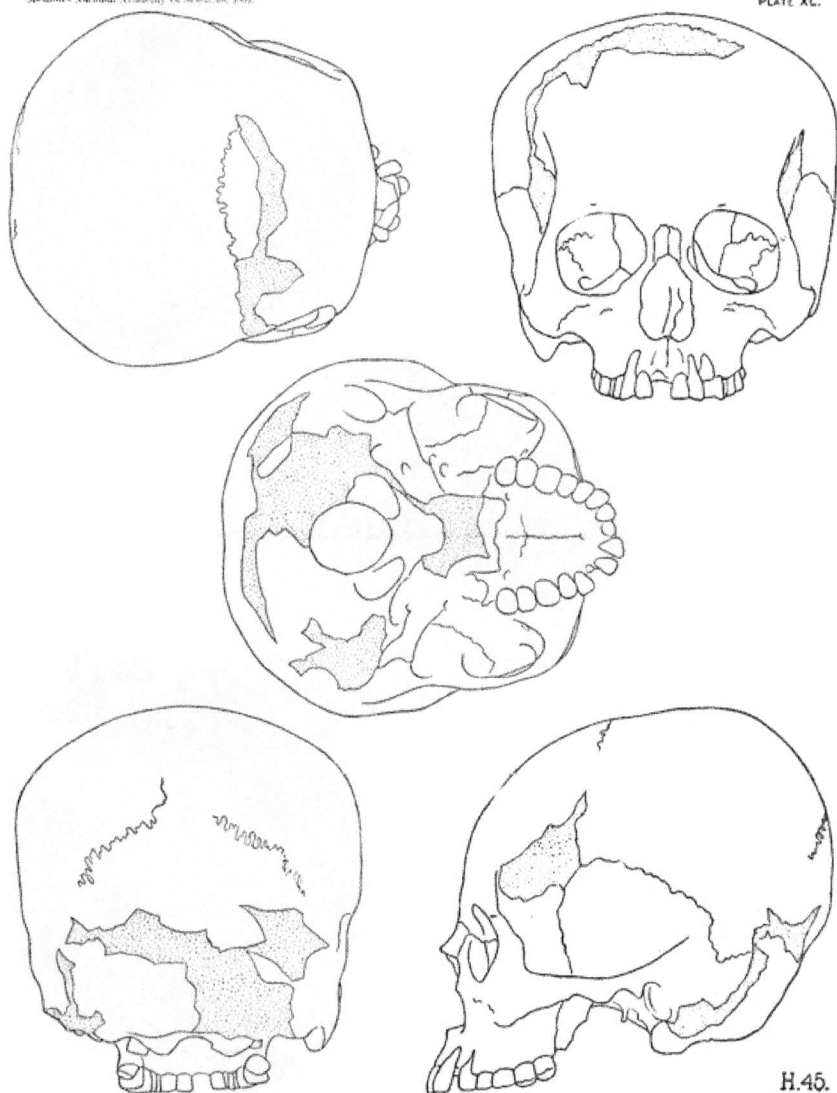

H.45.

PLATE XLI.

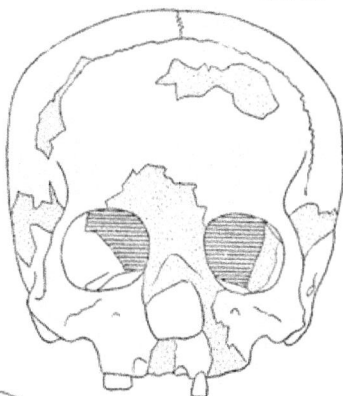

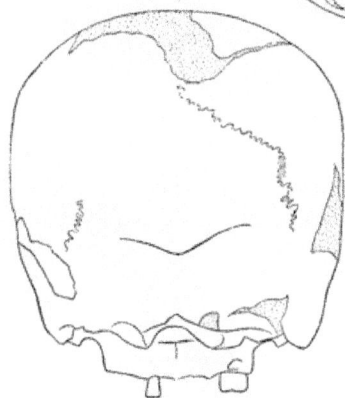

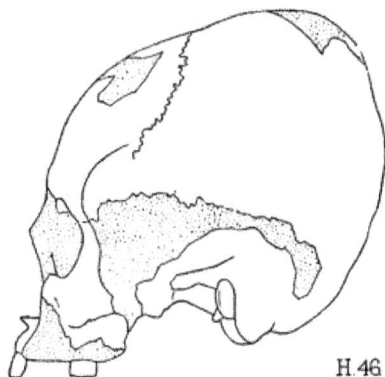

H.46

PLATE XLII.

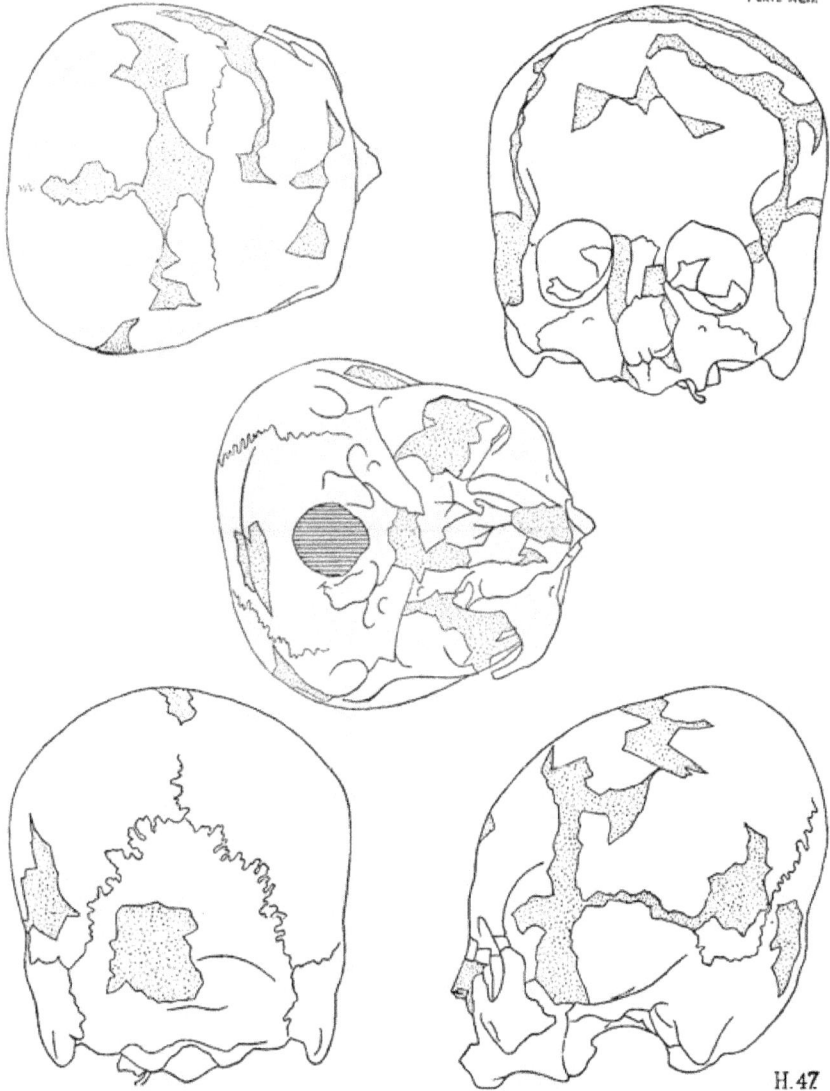

H.47

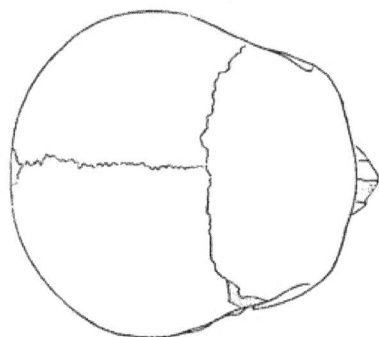

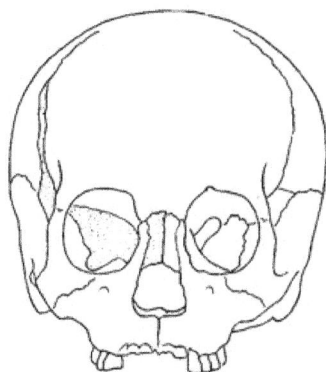

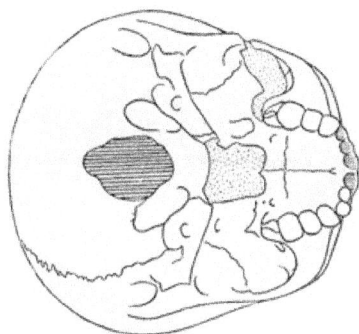

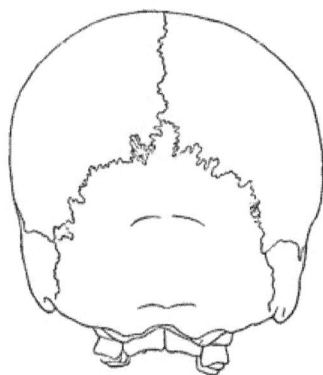

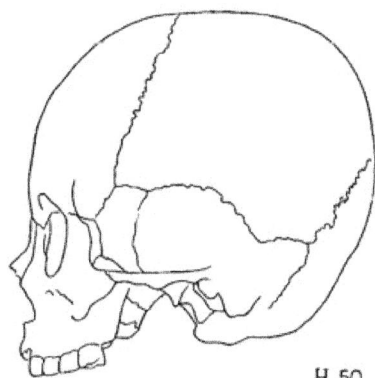

PLATE XLIII.

H 50

PLATE XLIV.

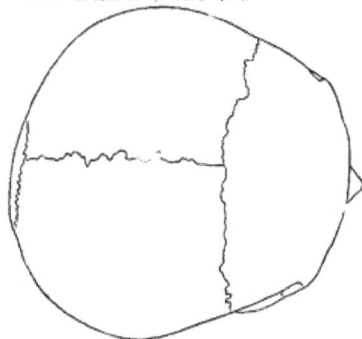

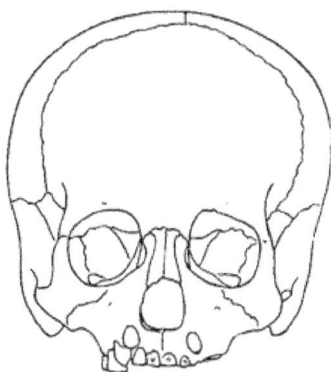

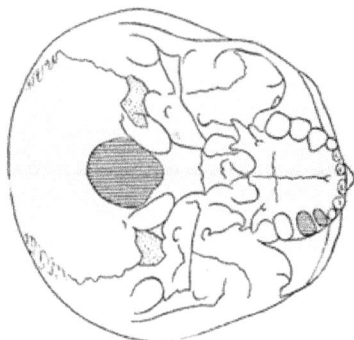

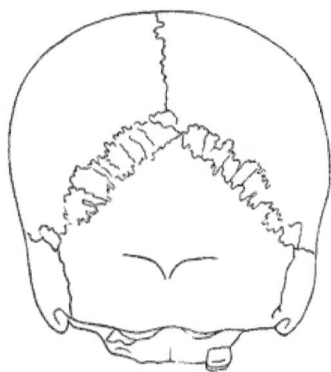

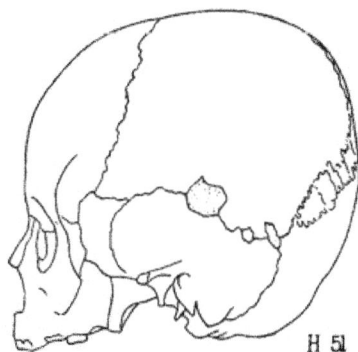

H 51

PLATE XLV.

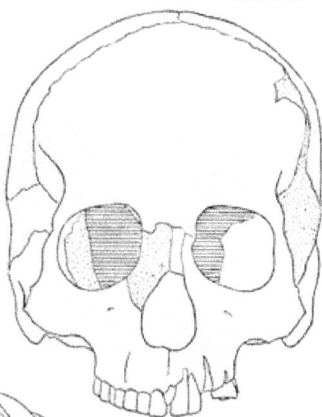

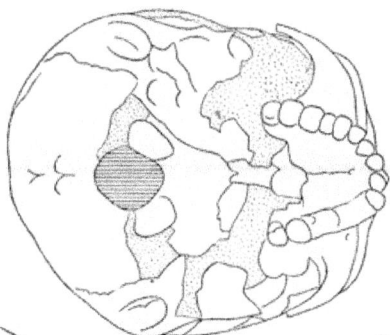

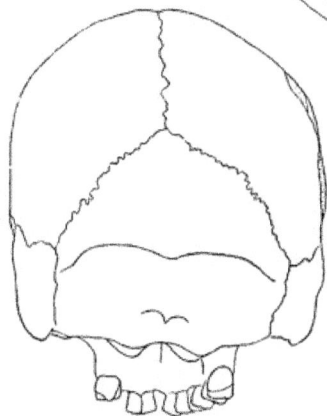

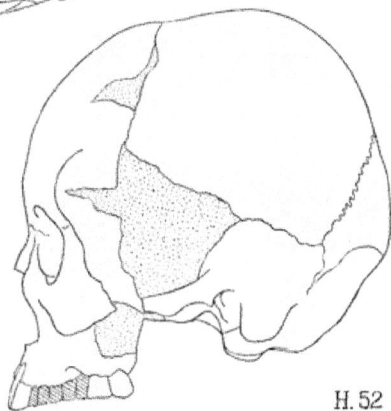

H. 52

PLATE XLVI.

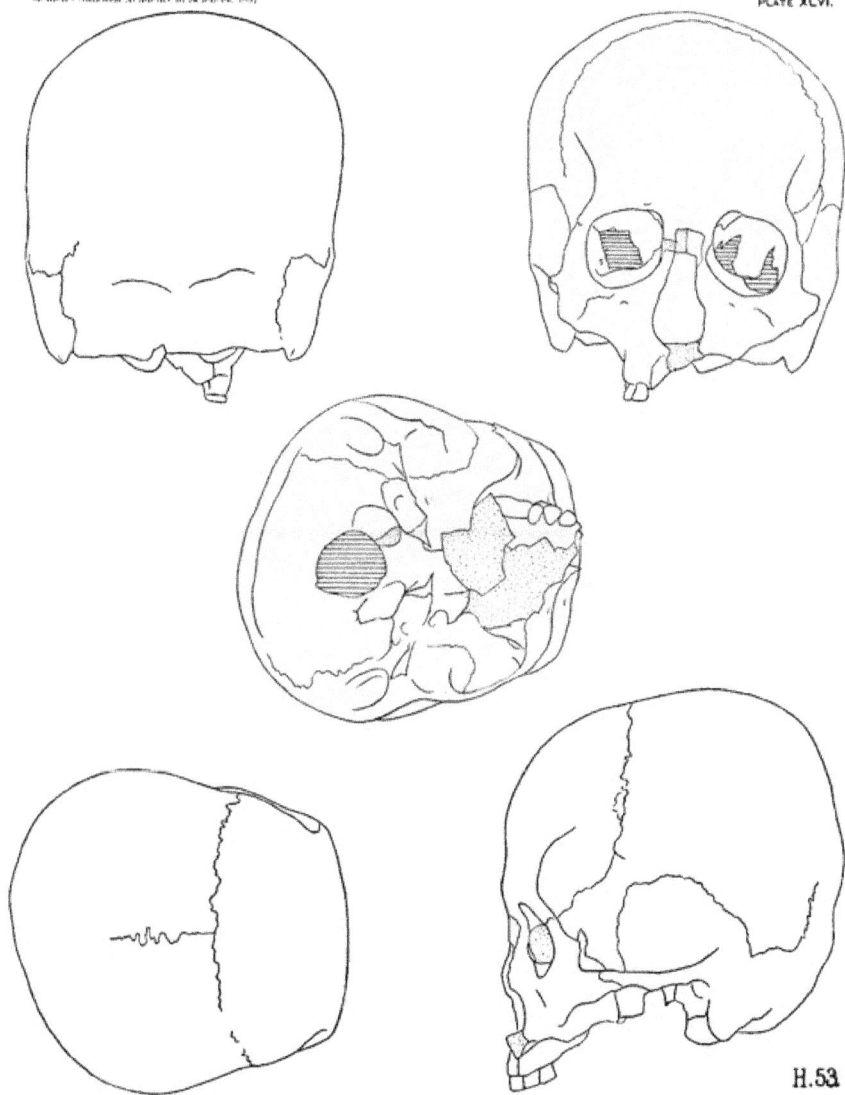

H.53

PLATE XLVII.

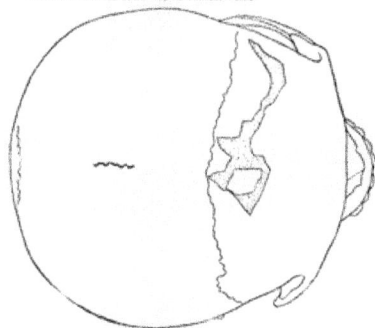
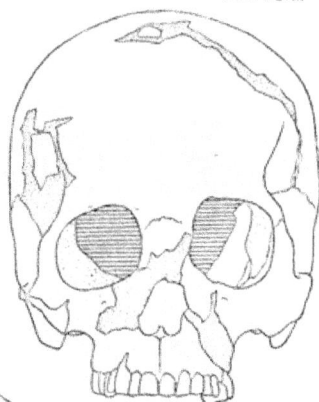
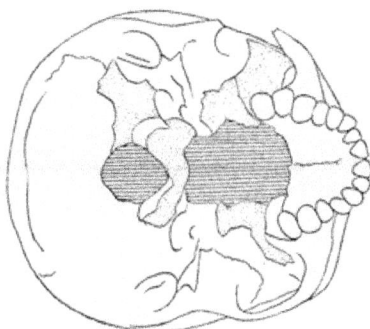
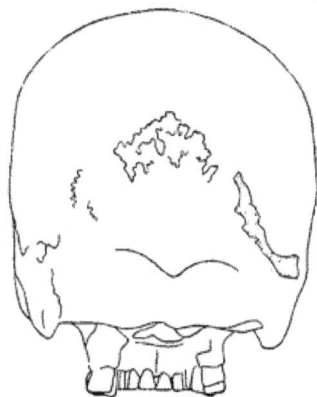

H 54.

PLATE XLVIII.

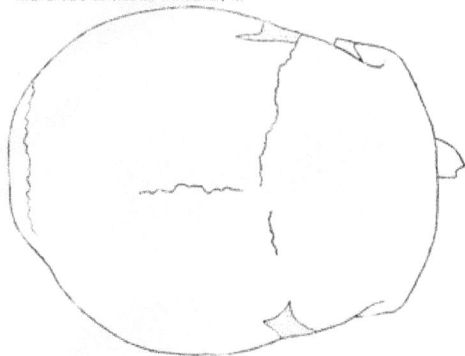

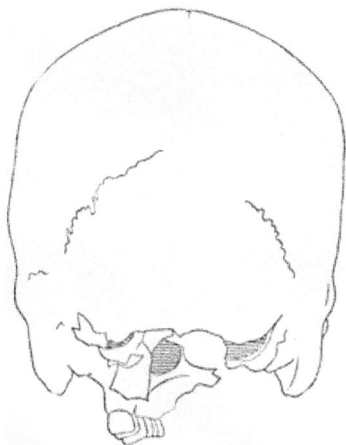

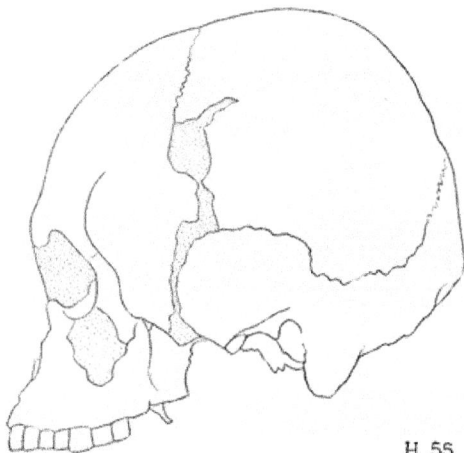

H. 55.

PLATE XLIX.

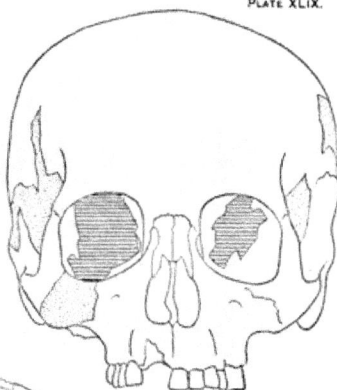

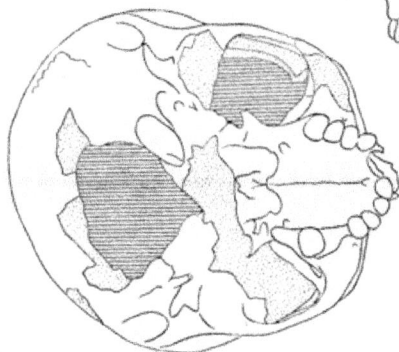

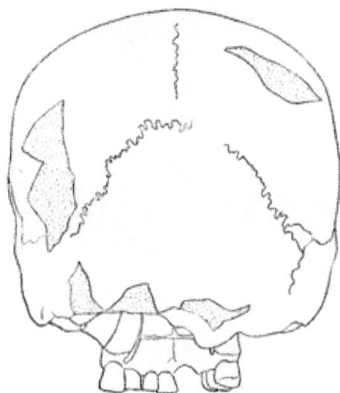

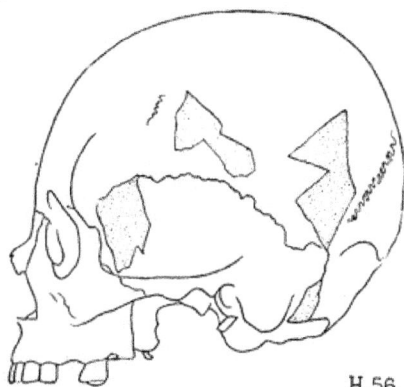

H.56

PLATE L.

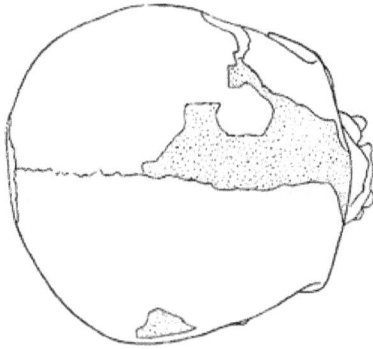

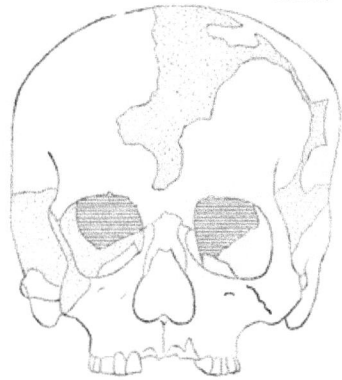

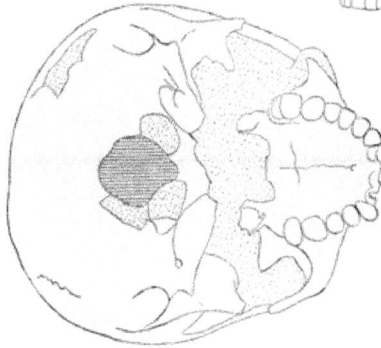

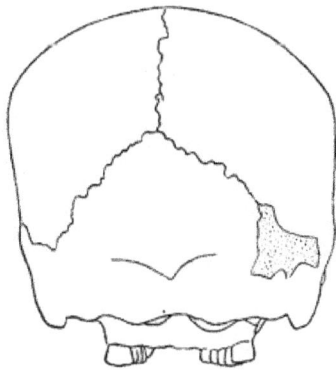

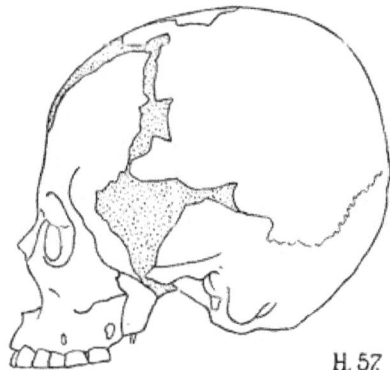

H. 57.

PLATE LI.

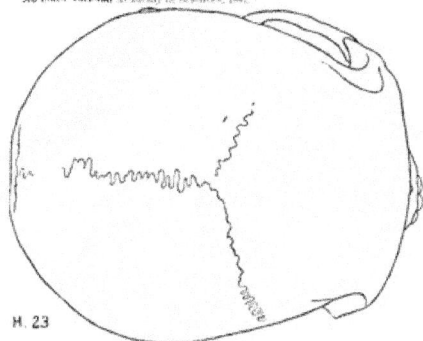

H. 23

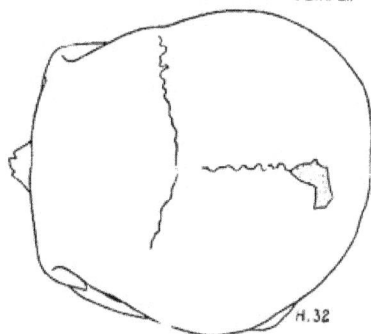

H. 32

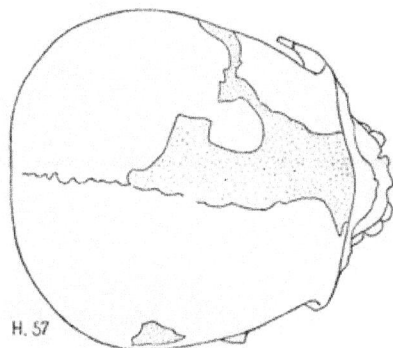

H. 57

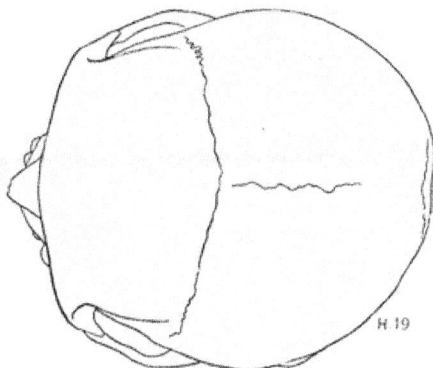

H. 19

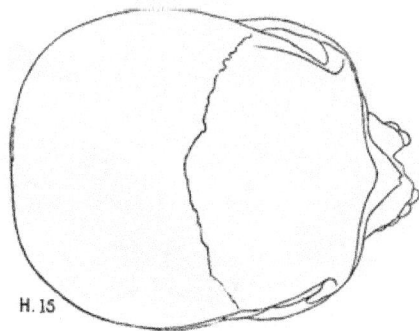

H. 15

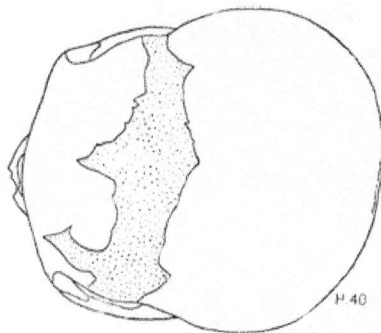

H. 40

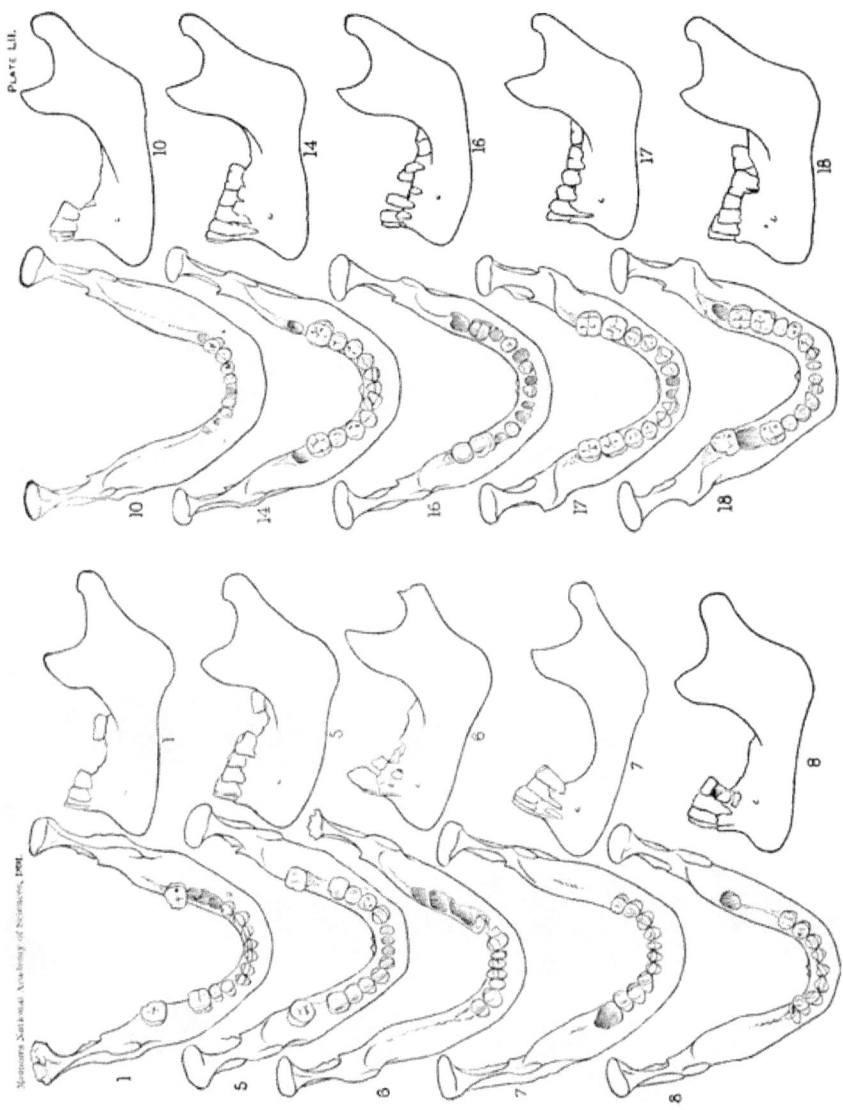

PLATE LII.

PLATE LIV.

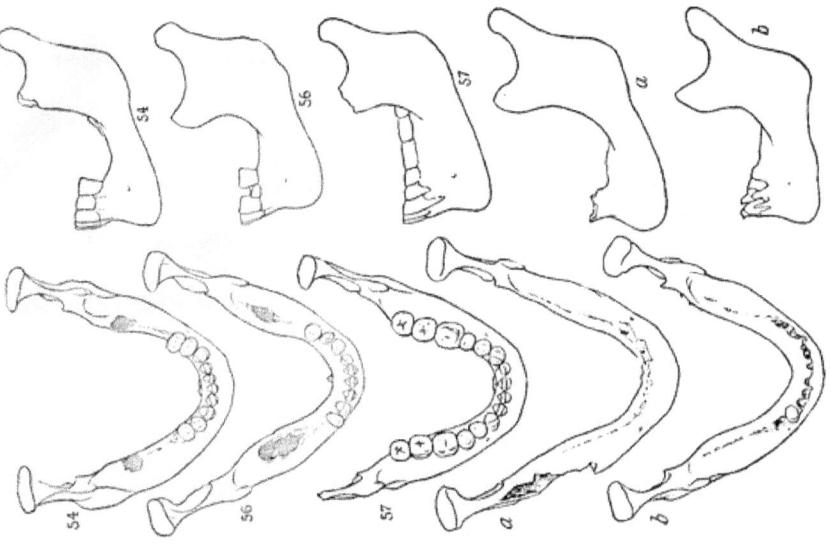

54 56 57 a b

54 56 57 a b

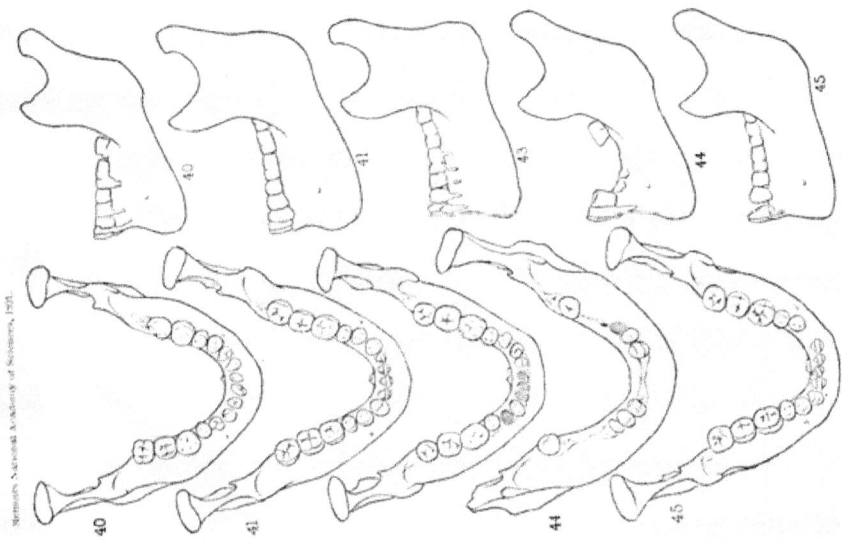

40 41 41 43 44 45

40 41 41 44 45

Memoirs National Academy of Sciences, 1921.

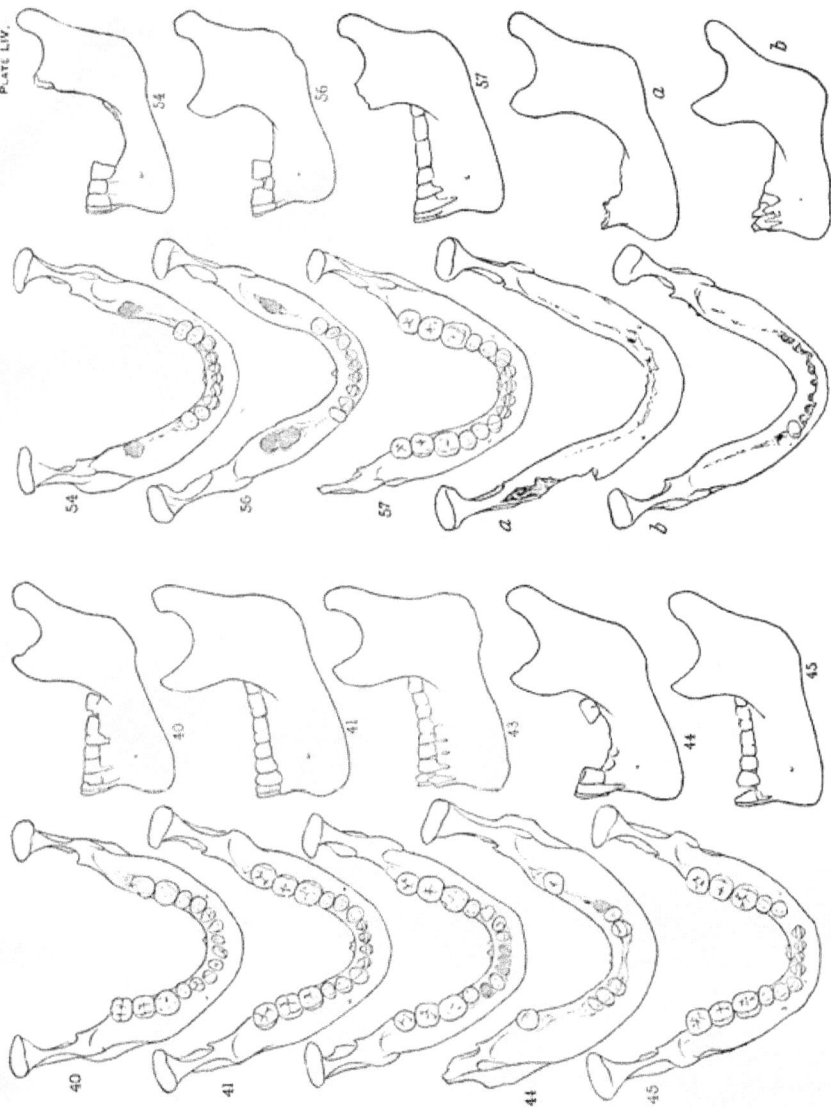

PLATE LIV.

54 56 57 a b

40 41 43 44 45

PLATE LVI.

Memoirs National Academy of Sciences. 1921.

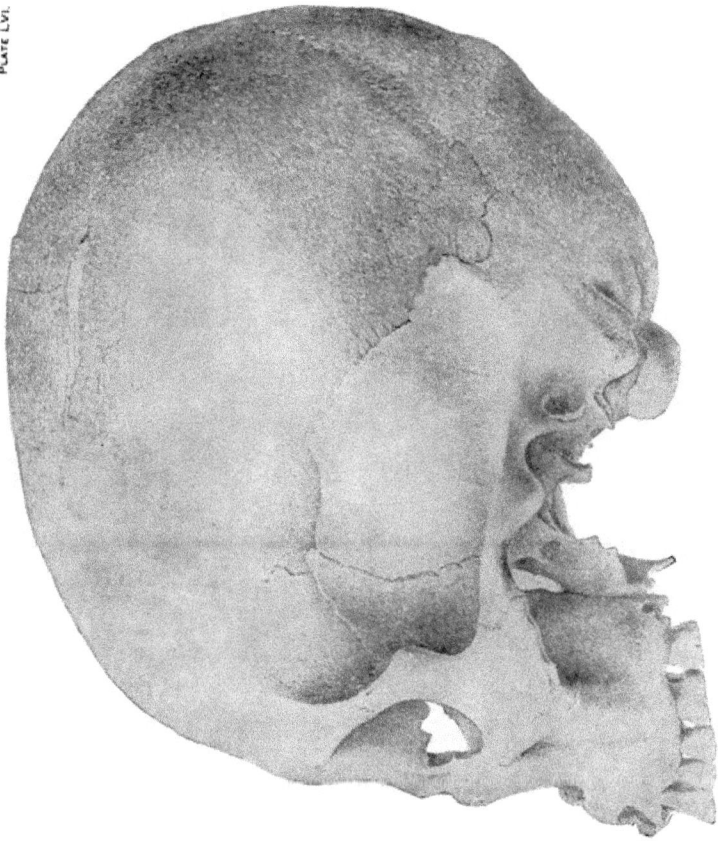

H No 40

PLATE LVII.

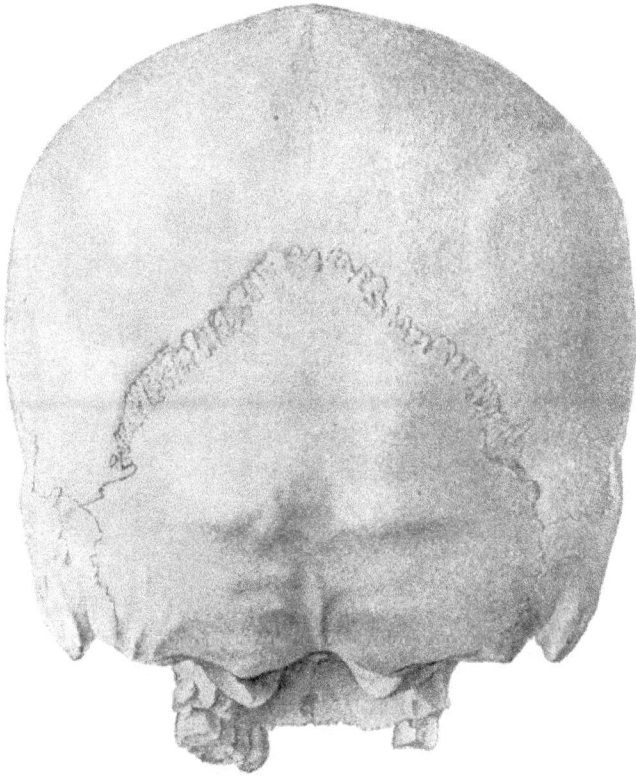

H No 40.

PLATE LVIII.

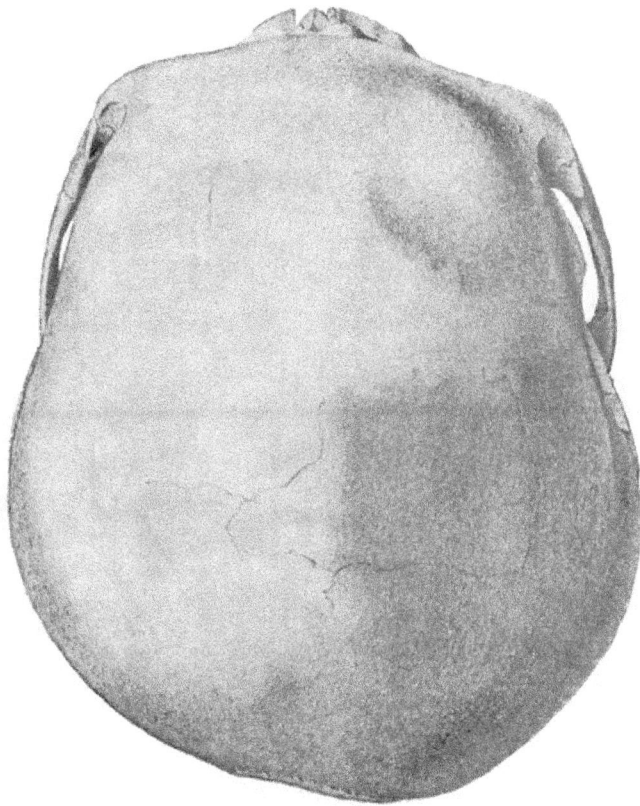

H. No 40.

PLATE LIX.

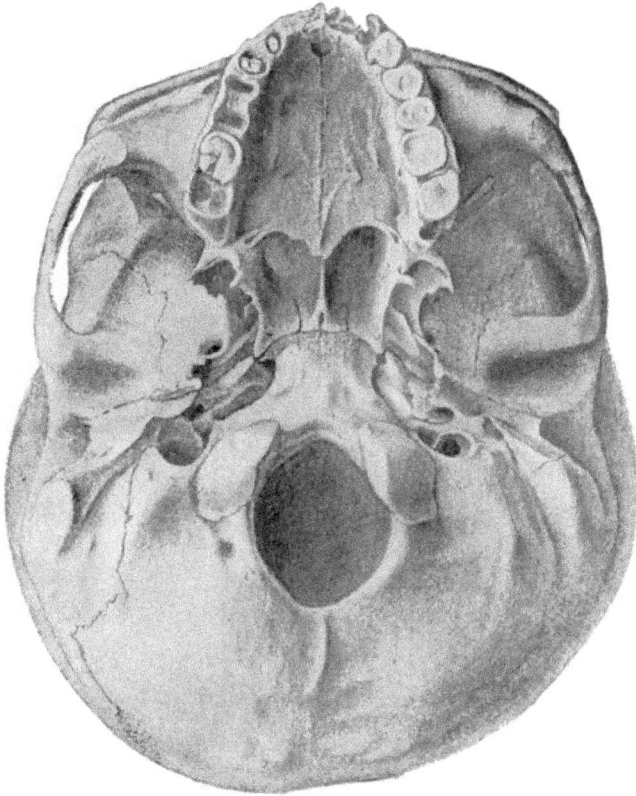

H No. 40.